GIS and BIM in Infrastructure Projects

GIS and BIM in Infrastructure Projects: An Introduction to Applications Across the Lifecycle demonstrates geographic information system (GIS) can complement building information modeling (BIM) throughout infrastructure development. While BIM focuses on geometrical and internal functional characteristics, GIS provides an essential wider locational context and spatial awareness, creating comprehensive solutions for infrastructure lifecycle management. The book emphasizes that infrastructure lifestyle planning cannot be done in isolation but must consider surrounding facilities, utilities, neighborhoods, and topography.

Structured around nine chapters covering all lifecycle stages of infrastructure projects. After introducing GIS fundamentals, the text explores specific applications across six lifecycle stages: feasibility (database management, site selection, environmental impact assessment), design (public participation, cost estimation, sunlight visualization), construction planning (scheduling, 4D modeling, safety planning), construction (contractor screening, site layout, materials management), operation and maintenance (infrastructure resilience, energy performance), and demolition (waste management, environmental impacts). Each chapter begins with learning objectives and concludes with summaries, supported by high-quality figures and illustrations. The final chapter identifies future research directions across all stages.

This is an ideal textbook for graduate and undergraduate students in architecture, engineering, and construction programs, as well as a useful book for AEC industry professionals seeking to implement GIS and BIM technologies in combination. It serves as a useful classroom resource and self-study manual, addressing the growing need for integrated BIM-GIS solutions in sustainable infrastructure development and lifecycle management.

Vijay Kr. Bansal is a faculty member of the Department of Civil Engineering at the National Institute of Technology (NIT), Hamirpur, India. He has more than two and a half decades of experience in teaching and research. He is the author of the textbook titled: *Project Management: Planning and Scheduling Techniques*, published by Routledge: Taylor & Francis. He is also a resource person for many international e-learning hubs. He has guided many PhD and Master's Programme theses and completed many research projects. His main research interests include application areas of geographic information system (GIS) and building information modeling (BIM) in the architecture, engineering, and construction (AEC) Industry. His other research interests are GIS-based 4D scheduling, repetitive scheduling, and location-based planning and scheduling. He has contributed many research papers to national and international journals.

GIS and BIM in Infrastructure Projects

An Introduction to Applications Across the Lifecycle

Vijay Kr. Bansal

Routledge
Taylor & Francis Group
LONDON AND NEW YORK

Designed cover image: © Retired Col. Vineet Singh

First published 2026
by Routledge
4 Park Square, Milton Park, Abingdon, Oxon OX14 4RN

and by Routledge
605 Third Avenue, New York, NY 10158

Routledge is an imprint of the Taylor & Francis Group, an informa business

British Library Cataloguing-in-Publication Data
A catalogue record for this book is available from the British Library

ISBN: 9781041284185 (hbk)
ISBN: 9781041281962 (pbk)
ISBN: 9781003759348 (ebk)

DOI: 10.1201/9781003759348

Typeset in Times New Roman
by KnowledgeWorks Global Ltd.

Contents

Preface

Infrastructure development is about making the world work to support the nations. The professionals of the AEC industry work together throughout the lifecycle of infrastructure projects. Often, various operations in the different lifecycle stages of infrastructure projects in the AEC industry are performed using conventional methods. However, the industry is going through a significant shift from the use of conventional methods to automation through the use of digital technologies after their invention. BIM is one of the widely used digital technologies in the different lifecycle stages of infrastructure projects.

An infrastructure is an important component in its surroundings for its entire lifecycle. Therefore, the lifecycle of an infrastructure must not be seen in isolation but in close relation with respect to all nearby existing facilities and utilities, the immediate neighborhood, and topography. Hence, the lifecycle of an infrastructure must be planned by keeping in view the site conditions, site constraints, and relationships among the already existing and the proposed infrastructure. The addition of a new infrastructure affects its surroundings, and the surroundings affect an infrastructure throughout its lifecycle. Therefore, locational aspects of an infrastructure are equally important along with its geometrical and internal functional characteristics. The planning and management of the lifecycle of an infrastructure project using BIM is generally in isolation, that is, without considering the participation of the infrastructure in its surroundings. BIM mainly focuses on the representation of physical dimensions, shapes, material properties, spatial relationships, and internal functional characteristics, such as mechanical, electrical, and plumbing systems of an infrastructure. BIM models how heat, ventilation, and air conditioning ducts, water supply pipelines, and electrical conduits are arranged inside walls and ceilings, ensuring that the infrastructure functions efficiently once built. On the other hand, the GIS is used to consider the locational aspects of infrastructure.

The planning and management of the lifecycles of infrastructure projects at different geographical locations have different sets of challenges, for which the use of GIS offers innovative solutions. Keeping in view the importance of the geographical location of an infrastructure as well as its geometrical and internal functional characteristics, BIM alone may not provide the complete solution for the lifecycle management of an infrastructure. GIS is supplementing the application areas in which BIM software tools are lacking. GIS allows an infrastructure modeled in BIM to have better geospatial context and virtual awareness of its surroundings. Hence, BIM and GIS together provide better solutions for the planning and management of the lifecycle of an infrastructure project.

The planning and management of the lifecycle of an infrastructure are linked to the utility networks in its surroundings, which provide water, electricity, fuel, sanitation, resources, communications, etc. The interaction among existing facilities and the proposed infrastructure and environmental impacts during the entire lifecycle are important aspects to be considered for coordinated, integrated, and sustainable infrastructure planning and management. It is like the judgment from the mono-disciplinary to the multi-disciplinary perspective. The use of BIM and GIS in the infrastructure lifecycle planning and management provides a multidisciplinary approach for collaborative and sustainable solutions. This strongly generates the need for the use of BIM and GIS in the AEC industry for planning and management of the lifecycle of infrastructure projects.

Understanding the difference between planning and management of the lifecycle of infrastructure projects, when done in isolation using BIM (generally without considering surroundings) and when done with respect to its surroundings, using BIM and GIS, to have a coordinated and integrated solution is very significant. This is like planning and management the lifecycle in a single dimension versus n-dimensions. It is the convergence of multi-disciplinary professionals together in contrast to the current fragmented nature of the existing lifecycle planning practices. Further, the ever-changing priorities of sustainable development and minimizing environmental impacts increase complexity in the lifecycle planning, which generates the need to use both BIM and GIS.

GIS is the technology that is generally not associated with the AEC industry; therefore, it is a little difficult to implement immediately in the AEC industry. Several GIS-based software tools are suggested in the literature for the industry, and their applicability on infrastructure projects is demonstrated through a few case studies; however, their routine use on real-life projects is rare. The professionals involved in the lifecycle stages of infrastructure projects do not find it time and cost-effective to invest in the GIS. Hence, a holistic approach is required where GIS is conceived, redesigned, and used in all lifecycle stages of infrastructure projects. The author felt the need for a simple book to make the AEC industry professionals aware of the application areas of GIS and BIM in the different lifecycle stages of infrastructure projects. Keeping this requirement in view, the idea of writing this book has been developed.

The major contribution of the book is threefold. First, a brief introduction to the basic concepts of GIS is provided in Chapter 2. Second, a discussion on the application areas of GIS and BIM in the six lifecycle stages of infrastructure projects in Chapters 3 to 8. The primary emphasis of this book is on the application areas of GIS in different lifecycle stages, as numerous books and literature already cover the applications of BIM extensively. Chapter 3 covers the application areas of GIS and BIM in the feasibility stage. The GIS application areas include database management, modeling of surroundings, modeling of surface and sub-surface facilities or utilities, modeling of site conditions and constraints, modeling of spatial relationships, selection of a sustainable site, planning site access, extending services to a site, etc. Chapter 4 covers the application areas of GIS and BIM in the design stage. The GIS application areas include public participation in design, quantity takeoffs, cost estimation, sunlight visualization, etc. Chapter 5 covers the application areas of GIS and BIM in the construction planning stage. The GIS application areas include calculation of site-specific rates of different items of work, planning and scheduling of repetitive and non-repetitive infrastructure construction projects, location-based planning and scheduling, 4D modeling, resolving workspace conflicts, planning for safety during construction, complex visualization, etc. Chapter 6 covers the application areas of GIS and BIM in the construction stage. The GIS application areas include screening of contractors, planning for the human workforce, materials trading, site layout planning, earthwork planning, site drainage planning, identification of locations for heavy equipment on a site, reducing construction wastage, tracking site dynamics, etc. Chapter 7 covers the application areas of GIS and BIM in the operation and maintenance stage. The GIS application areas include site drainage planning, making maintenance decisions, inspection and safety, energy performance, environmental impact assessment, infrastructure operation, etc. Chapter 8 covers the application areas of GIS and BIM in the demolition stage. The GIS application areas include demolition waste management, carbon emission, environmental impact assessment, emergency management, damage assessment, etc. Third, the application areas of GIS in the different lifecycle stages of infrastructure projects that need future research are highlighted. The book helps in understanding the landscape of work done to use GIS and BIM in the lifecycle stages of infrastructure projects.

The book addresses the readers who aspire to use GIS and BIM in the AEC industry. The book may be used for both undergraduate and graduate courses in architecture, engineering, and construction. Each chapter starts with learning objectives and ends with a chapter summary. The book is an easy way for the AEC industry professionals to get easy self-study materials on the application areas of GIS and BIM in the different lifecycle stages of infrastructure projects. The author has illustrated application areas with theory and diagrams where necessary for developing clear, understandable, and actionable content for the AEC industry professionals.

The author hopes that he has succeeded in the writing part; however, the author is sure that students, teachers, and the AEC industry professionals will have suggestions, corrections, and criticisms on the text presented in the book. The author encourages readers to send their valuable suggestions, corrections, and criticisms so that the author can include necessary changes in future editions. Please mail your comments to:

Vijay Kr. Bansal
Email: vkb@nith.ac.in
Website: https://nith.ac.in/

Chapter 1

Need for GIS in the AEC Industry

1.1 Learning Objectives

After the completion of this chapter, readers will be able to understand the:

- use of building information modeling (BIM) in the planning and management of the lifecycle of infrastructure projects,
- need for the geographic information system (GIS) in the lifecycle of infrastructure projects, and
- need for integration of BIM and GIS in the planning and management of infrastructure lifecycle.

1.2 Introduction

Infrastructure development is about making the world work to support society and the nations. The construction of buildings, roads, dams, bridges, public utilities, and other forms of infrastructure are significant components in the development of a nation. The architecture, engineering, and construction (AEC) industry is the collaboration of architects, engineers, and construction professionals to ensure smooth planning, construction, operation, and maintenance of residential, commercial, or industrial infrastructure projects. AEC professionals work together throughout the lifecycle stages of infrastructure projects. Architects design an infrastructure project for construction or restoration. Architects work to make the design aesthetically pleasing, feasible, executable, and sustainable while ensuring compliance with local bylaws, codes, and standards. Engineers take care of structural design, stability, execution, construction, operation, and maintenance of infrastructure projects. Engineers make sure that the project is technically feasible, stable, executable, and complies with government regulations. The construction professionals are responsible for the construction of infrastructure projects, and for this, a plan and an execution schedule are developed, all arrangements for deciding a contractor are made, construction site risks are mitigated, and construction progress as per the plan and schedule is ensured. During the project execution, construction professionals communicate with architects and engineers to ensure that the infrastructure project is being executed as per the plan, schedule, agreed quality, safety norms, and all legalities are taken care of.

DOI: 10.1201/9781003759348-1

Sometimes, many of the operations in the different lifecycle stages of infrastructure projects in the AEC industry are carried out using conventional methods. However, the AEC industry is undergoing a significant shift from the use of conventional methods to automation through the use of digital technologies after their inventions. Digital technologies, in general, refer to information and communication technologies to facilitate information development, various types of analyses, data handling, and storage in digital form for promoting effective communication among AEC professionals. The widely used information and communication technologies in infrastructure projects are building information modeling (BIM), augmented reality, radio frequency identification devices, remote sensing (RS), geographic information system (GIS), global positioning systems (GPS), quick response codes, robotics, and artificial intelligence.

BIM is one of the widely used information and communication technologies for infrastructure project documentation, work collaboration, sharing project information, etc., throughout the lifecycle of a project. BIM software tools are used by individuals, businesses, and government agencies for planning, design, construction, operation, and maintenance of buildings and diverse physical infrastructures such as water supply, electricity, gas, communication utilities, roads, railways, bridges, ports, tunnels, etc. Augmented reality is used for construction planning, personal training, decision-making, reporting real-time project progress, etc. Radio frequency identification devices are used for maintaining transparency and enhancing the productivity of the delivery supply chain in the AEC industry.

RS involves the use of sensors mounted on low or high-flying aircraft to collect data about the surface of the earth, including information about the topography, land use and land cover (LULC), and other factors that may affect an infrastructure project. GIS is used for site-specific data analysis, terrain analysis, geospatial analysis, etc. GPS give the real-time location of an object that is used for tracking site dynamics, site monitoring, etc., in the lifecycle of infrastructure projects. Quick response codes provide information quickly about locations of site consignments, equipment, materials, and even workers. Mobile and laser scanning devices are used for reading quick response codes in the AEC industry. Robotics are used in the industry for the automation of different operations and for increasing quality, progress, and safety during project execution. Artificial intelligence is also used for improving efficiency, accurate analysis of large data volumes, etc., in the AEC industry. This chapter mainly focuses on the discussion behind the use of BIM in the AEC industry and why GIS and integration of BIM and GIS are needed.

1.3 BIM in Infrastructure Lifecycle

Infrastructure lifecycle management involves various stages in which an infrastructure is planned, designed, constructed, and maintained. In the AEC industry, infrastructure lifecycle management is generally divided into six stages: *feasibility*,

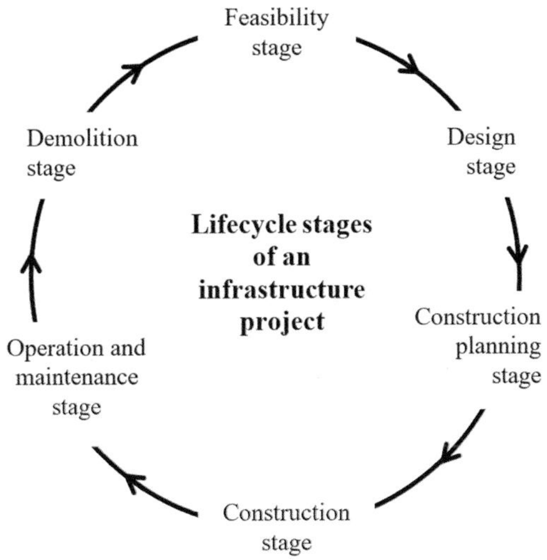

Figure 1.1 Lifecycle stages of an infrastructure project

design, *planning*, *construction*, *operation and maintenance* (O&M), and *demolition,* as shown in Figure 1.1 (Guo et al. 2010). Each lifecycle stage has different functions, and all stages are linked to each other for information flow. Infrastructure lifecycle management involves managing the entire lifecycle of an infrastructure from *feasibility* to *demolition* stage for information sharing, smooth work execution, coordination among stakeholders at various stages, and environmental sustainability.

Infrastructure lifecycle planning and management make use of information which are generated and used in the various stages of the lifecycle. Information is communicated in the forms: (i) location maps, contour maps, detail project reports, all best possible alternatives, etc., in the *feasibility* stage; (ii) access plans, two dimensional (2D) drawings, three dimensional (3D) models, elevations, cross-sectional views, structural design, specifications, various items of work involved, cost estimates, etc., in the *design* stage; (iii) execution schedule, materials schedule, equipment schedule, labor schedule, financial schedule, site layouts, etc., in the *planning* stage; (iv) tender-documents, components delivery schedule, safety and quality control recommendations, as build-drawings/records, etc., in the *construction* stage; (v) operation manuals, maintenance plan, maintenance records, etc., in the O&M stage; and (vi) records generated during demolition planning, etc., in the demolition stage. The sharing of such information among various stages of the lifecycle of infrastructure projects is an essential requirement. Infrastructure lifecycle planning and management rely heavily on information platforms for maintaining and sharing information.

The information and communication technologies have been widely regarded as the key enablers for performance improvement in the AEC industry, as discussed in the last section. The software tools for various disciplines like architecture, structure, construction, engineering, etc., are being used to maintain and analyze the information throughout the lifecycle of an infrastructure project. The drafting group generally uses drafting tools, a planner uses project planning software tools, whereas the construction and operations group uses blueprints or printed paper, etc. Bridging their outputs for effective use is a major hurdle in the AEC industry. Further, the information flow between various groups working in different stages of an infrastructure project is not through well-defined standards and procedures. This results in data redundancy, redundant processes, and poor information quality. An infrastructure project has information in several graphical (drawings, blueprints, etc.) and non-graphical (schedule, cost estimate, specifications, etc.) forms that are developed, maintained, and analyzed separately by different groups of people in different stages of an infrastructure project with the help of different commercially available software tools.

The professionals working in the AEC industry have to go through this information, which sometimes does not agree with each other. The overlaps and lack of consistency in the information often lead to errors, which result in miscommunications that are expensive and time-consuming in nature. BIM is used as a practical information-sharing platform nowadays during the entire lifecycle of infrastructure projects. The information generated and used in different lifecycle stages of an infrastructure project without BIM is available in different information formats generated using various tools/software. This leads to interoperability problems because of the difficulty in bridging various types of information formats obtained by using various tools/software. BIM resolves interoperability problems throughout the lifecycle of an infrastructure project (Ozturk 2020; Underwood and Isikdag 2010).

Poor and inefficient sharing of information among owners, architects, engineers, users, and the public has always affected the AEC industry. This lack of effective communication, coördination, and data sharing has a direct impact on claims, public relations, lifecycle costs, etc. If information is maintained in a single BIM platform, changes are made to these documents at one place only; drawings, schedules, cost estimates, and specifications will be consistent with each other. This change significantly reduces the communication problems that often slow down the project's progress and increase the costs. BIM is used in almost every lifecycle stage of an infrastructure project (Jiao et al. 2013). For the professionals involved in an infrastructure project, BIM enables the use of the virtual information model that is shared among the design team (architects, landscape planners, surveyors, civil engineers, structural and building services engineers, etc.), main and sub-contractors, and the owner/operator. Each professional adds discipline-specific information to the virtual model, and it becomes the product of several different disciplines that worked together. It goes beyond the planning and design stages of an infrastructure project, useful for the entire lifecycle. The concept of sustainability in the infrastructure lifecycle management from the BIM viewpoint is also being explored.

Out of the several information and communication technologies, BIM is one of the most promising developments in the AEC industry. BIM represents a new paradigm that encourages the integration of stakeholders from different lifecycle stages of an infrastructure project. An accurate 3D virtual model is digitally constructed in BIM that characterizes the geometrical characteristics of an infrastructure. It helps AEC professionals in visualizing what is to be developed in a virtual environment to identify any potential design, construction, or operational issues. The 3D model serves as a centralized repository for all the stakeholders involved to put in and pull out the necessary information. In addition to this, BIM is used for 4D modeling by linking 3D components of an infrastructure with the execution schedule. Linking construction costs with the components makes it a 5D model. BIM provides a 6D model to answer all the questions, like who, what, when, and how, in the different stages of an infrastructure project lifecycle.

A 3D virtual model developed using BIM is used for visualization, modeling of spatial relationships, development of fabrication/shop drawings, code reviews, and modeling of rules and regulations with respect to local bylaws, construction safety, fire safety, etc. A 3D virtual model is used for the estimation of quantities of various items of work. The materials quantities required for each item of work are automatically extracted from the 3D model and updated for every change in the 3D model. BIM helps in sequencing construction (preparation of execution schedule, materials requirement schedule, labor requirement schedule, equipment requirement schedule, and delivery schedule of infrastructure components), detection of collision among components (like piping intersects with steel beams, ducts, or walls), identifying potential failures, development of evacuation plans, and so forth. BIM applications are also reported for renovations, space planning, maintenance operations, and sustainable design. The infrastructure proposals are analyzed, simulated, and performance is checked with the help of BIM. Its application areas are not limited to the design and construction stage of an infrastructure, but throughout the lifecycle. Information about an element of an infrastructure in BIM is used throughout its lifecycle.

1.4 Need of GIS for the AEC Industry

The lifecycle planning and management of an infrastructure project depend upon the level of experience, knowledge, individual perspective, and how a professional understands the surroundings of an infrastructure. Any infrastructure, either big or small, is an important component in its surroundings. It cannot remain in isolation but is closely related to all other existing facilities and utilities in its surroundings. Just as a single entity, any new infrastructure addition creates many changes in its surroundings. Hence, the lifecycle of an infrastructure must be planned by keeping in view the site conditions, site constraints, and relationships among the proposed infrastructure and its surroundings. The lifecycle of an infrastructure, especially in hill regions where topography plays a major role, cannot be planned and managed without geospatial modeling and analysis capabilities, which are available in GIS.

An infrastructure is an important monument for its entire lifecycle in its neighborhood. Hence, its lifecycle must not be seen in isolation but in close relation with respect to all nearby existing facilities and utilities, neighborhoods, and topography. The addition of an infrastructure affects its surroundings or surroundings are affected by the addition of an infrastructure throughout the lifecycle. The locational aspects of an infrastructure are equally important, along with its geometrical and internal functional characteristics throughout the lifecycle. Infrastructure lifecycle planning and management using BIM is generally in isolation, that is, without considering the participation of an infrastructure in its surroundings. BIM mainly focuses on the representation of physical dimensions, shapes, material properties, spatial relationships, and internal functional characteristics, such as mechanical, electrical, and plumbing systems of an infrastructure. BIM models how heat, ventilation and air conditioning ducts, water supply pipelines, and electrical conduits are arranged inside walls and ceilings, ensuring that the infrastructure functions efficiently once built. On the other hand, the GIS is used to consider the locational aspects of infrastructure (Bansal 2017).

GIS is being used widely in the different lifecycle stages of an infrastructure project by a wide range of specialists to solve different problems. It is a computer-based tool used extensively to solve various AEC industry problems involving the use of both graphical and non-graphical information. GIS is supplementing the already existing abilities and knowledge in the AEC industry. It is a tool for creating, managing, analyzing, and visualizing all types of geographical or locational information. GIS provides a medium to interact with relevant locational information so that experts and non-experts are able to visualize the environmental consequences of a big infrastructure project addition (Bowman 1998).

Lifecycle planning and management of infrastructure are linked to nearby existing facilities and utilities in the surroundings, which provide water, electricity, fuel, resources, communications, etc. The layouts of such facilities and utilities are now GIS-based in many countries. Many cities have GIS-based information available that typically includes site topography, LULC maps, existing utilities and facilities, footprints of existing infrastructure, street layouts, and demographics, which are valuable during infrastructure lifecycle management. The database management capability of GIS provides the mechanism for querying and retrieving graphical and non-graphical information to support infrastructure lifecycle management (Bansal 2016; Gunes and Kovel 2000). GIS also plays a major role in the planning of extensions of facilities and utilities like approaching roads, water supply, electricity, gas supply, sanitary sewers, etc., to the site of an infrastructure project. The survey and revenue records in the form of digital maps are imported into GIS. Attribute data of each parcel of land are populated with extensive information about the land owner, previous owners, area, price of acquisition, etc., in GIS. This digital database provides the required information as frequently as needed (Jeljeli et al. 1993).

The planning and management of infrastructure lifecycle in different geographical locations have different sets of challenges, in which the use of GIS has innovative solutions (Miles and Ho 1999). Keeping in view the importance of the geographical location of an infrastructure as well as its geometrical and internal functional characteristics, BIM alone may not be a complete solution for infrastructure lifecycle planning and management. GIS is supplementing the application areas in which BIM software tools are lacking. GIS allows an infrastructure modeled in BIM to have better geospatial context and awareness of its surroundings. The geospatial context for BIM means better lifecycle management.

The use of GIS has been reported in a wide range of studies for supporting various tasks of the infrastructure lifecycle. Understanding the participation of an infrastructure in its surroundings is made possible with the help of GIS. The layouts of different types of existing utility systems, like electricity, gas, water, sewage, etc., are now GIS-based which play a major role in site selection and space planning. The geospatial analysis with already available accurate and field-experimented data makes infrastructure lifecycle management faster and more accurate. GIS allows linking attributes with each component of an infrastructure that is used throughout the lifecycle. For instance, a wall represented in GIS may hold the attributes: physical characteristics (length, width, height, materials used, color, etc.), method and cost of construction, name of the contractor, sub-contractor constructed it, properties of materials used, past repair and maintenance records, etc.

In big infrastructure like hydroelectric power projects, GIS helps in the calculation of reservoir capacity, hydraulic head, costs of construction, site accessibility, environmental impacts, etc. Hydroelectric power projects have several components, like a reservoir, a dam, a powerhouse, a control room, a tail race tunnel, a head race tunnel, etc. The changes in the locations of these components and the corresponding impact on the power generation potential of a hydroelectric project are assessed using GIS.

1.5 BIM and GIS Are Living Separate Lives

The information in the form of text has been used widely to describe any object, process, or phenomenon taking place on the surface of the earth. In case, information in the form of text is not sufficient or cannot correctly describe complex objects, processes, or phenomena, some sketches or drawings accompany the text. Text and sketches/drawings together are frequently used in the AEC industry to describe objects on the surface of the Earth or the infrastructure to be constructed. People started using personal computers to manage text information with the invention of computers. Nowadays, professionals have very good computer-based tools to develop and manage text information which are in the form of text, numbers, tables, etc. Similarly, AEC professionals also use personal computers to create, edit, and manage text and sketches/drawings. Nowadays, AEC professionals have very good computer-aided drafting (CAD) software tools to deal with graphical information.

The developers who were developing and improving software tools for dealing with text information keep on enhancing the user-friendly capabilities of software tools for managing text information. Similarly, the developers who were developing and improving software tools to create, edit, or manage graphical information keep on adding user-friendly capabilities to software tools that manage graphical information. The AEC industry has seen an independent move of the developers working on both sides, as shown in Figure 1.2. The text information in the AEC industry is generally called non-graphical information; on the other hand, the information that contains sketches or drawings is called graphical information. The non-graphical information in the AEC industry generally includes reports, quantity takeoffs, cost estimates, execution schedules, specifications, code provisions, etc. The graphical information in the AEC industry generally includes site drawings, 2D plans, sections, views from different sides, working drawings, etc.

With time, professionals in the AEC industry developed and started using different software tools for different tasks. For example, drafting tools for making drawings, analysis and design tools for structural design, tools for cost estimation, tools for scheduling, etc. The use of different software tools by different professionals in the AEC industry for different tasks resulted in different information formats.

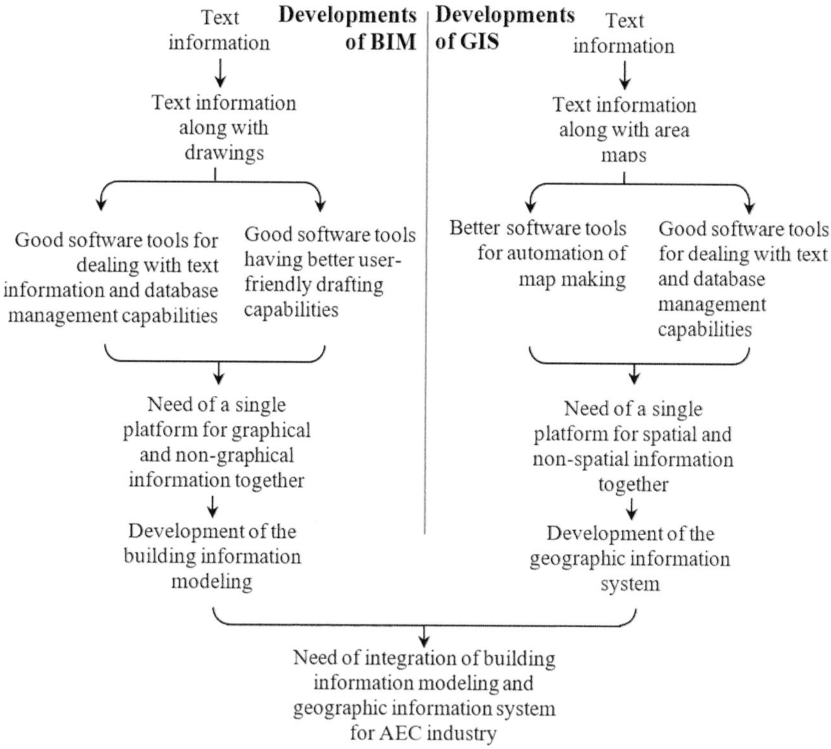

Figure 1.2 Need for integration of building information modeling and geographic information system for the AEC industry

The inputs and outputs of different software tools are difficult to bridge together for AEC industry professionals. This resulted in interoperability problems in the industry.

With time, the AEC industry realized that the use of multiple software tools for different tasks is consuming a lot of resources, energy, and increasing interoperability problems. This generated the need for a single platform for different tasks. This resulted in the development of BIM, a single tool/technology for different tasks as discussed in an earlier section. BIM handles both graphical and non-graphical information together on a single platform. It is possible to perform many tasks related to the lifecycle of an infrastructure in the AEC industry with the invention of BIM. Many tasks related to the lifecycle of an infrastructure on a single platform reduce the interoperability problems.

On the other side, in the field of geography, information in the form of text has been used to describe features, phenomena, or processes on the surface of the Earth. In cases where text information is insufficient or cannot correctly describe features, phenomena, or processes, maps accompanying the text are used. Text information and maps together are still widely used by geographers to describe features, phenomena, or processes on the surface of the Earth. Geographers started using personal computers to manage text information and maps, nowadays, good computer-based tools are used for map-making tasks.

The developers who were developing software tools to create, edit, or manage text information kept on adding good user-friendly capabilities to the software tools along with database management capabilities. Similarly, the developers who were developing software tools for map-making kept on enhancing user-friendly capabilities for map-making. The text information in the field of geography is called non-spatial, non-locational characteristics, or attributes; on the other hand, the information that contains maps or locations is called spatial or locational information. With time, geography professionals realized the need for a software tool that can deal with both spatial and non-spatial information together on a single platform. Along with the spatial and non-spatial information, professionals also felt the need for database management capabilities and geospatial analysis on a single platform. GIS provides a single platform to deal with spatial and non-spatial data on a single platform along with database management and geospatial analysis capabilities.

The lifecycle management of an infrastructure in a locality is not done in isolation, even a well-planned infrastructure construction cannot be executed properly if it is not planned with respect to its locality. In reality, any new infrastructure addition has an impact on the existing facilities; therefore, consideration must be given to the surroundings using GIS. The interaction between existing facilities and the proposed infrastructure and their environmental impact are important aspects to be considered for a coordinated, integrated, and sustainable infrastructure lifecycle.

The main difference that distinguishes the two technologies lies in the way as they represent entities in the space, how they tie these entities to a spatial reference, how they organize information, how they store information, and their capacity to perform analysis. BIM uses the Cartesian coordinate system to reference geometry to a local origin (0, 0, and 0) and does not typically reference the geometry to a location on the surface of the Earth. GIS uses the geographic coordinate system, either in the form of geographic coordinates (longitudes and latitudes) or a specific map projection with geographically referenced coordinates. Another difference between BIM and GIS is the way as they store spatial information. BIM supports a variety of primitives like cones, spheres, cylinders, freeform shapes, etc., while GIS supports mainly points, lines, and polygons. These barriers prevent an effective or full automation of planning and management of the lifecycle of an infrastructure for using the capabilities of both BIM and GIS.

1.6 Need for Integration of BIM and GIS

The use of BIM for 3D visualization, graphical rendering, documentation, modeling of spatial relationships, and compatibility with architectural design, sectional views, perspective views, and various plan views helps throughout the lifecycle of an infrastructure project. BIM is a 3D framework developed for 3D modeling, integrating complete graphical and non-graphical information corresponding to infrastructure components, and modeling of spatial relationships among various components in a single platform. The quantity takeoffs are easily done and updated if any change in the design is made to effectively ascertain potential costs (Underwood and Isikdag 2010). Keeping in view the importance of geometric as well as the locational information of an infrastructure project under consideration; BIM is used for maintaining the geometric information; on the other hand, GIS maintains the locational information of an infrastructure (Zlatanova and Prosperi 2005).

BIM is best suited for managing geometrical information required in the entire lifecycle of an infrastructure project; on the other hand, GIS is best suited for managing information about everything outside of an infrastructure. BIM enables the sharing of geometric information more easily to make more coordinated and informed decisions about an infrastructure; on the other hand, GIS enables more effective spatial decision-making at each stage of an infrastructure to ensure the sustainable lifecycle. The integration of BIM and GIS facilitates in understanding the effects of surroundings on each stage of an infrastructure lifecycle and impacts on surroundings because of infrastructure addition (Dore and Murphy 2012). BIM and GIS integration optimizes the materials use, reduces wastage, waste reuses, recycling to minimize waste materials, optimizes overall energy costs, minimizes demolition time, and reduces impacts on surroundings (Kim et al. 2011). The integration enables the use of diverse information datasets corresponding to an infrastructure and its surroundings together for a better understanding of the impacts of decisions during each stage of an infrastructure lifecycle (Kang and Hong 2015; Liu et al. 2017).

The major essence of using BIM and GIS together includes consideration of infrastructure geometry, site topography, immediate surroundings, interaction between infrastructure and its surroundings, and services of existing facilities/utilities in the lifecycle of an infrastructure project. This facilitates in getting the locations where utilities and facilities are and in understanding how water, sanitary sewer, and other facilities may be arranged on a site. This improves communication among stakeholders, provides geospatial analysis capability, and facilitates in understanding the impacts of an infrastructure on its surroundings. BIM and GIS have their own strengths and capabilities in their respective fields; one cannot replace the other with the present level of research to have a coordinated solution for planning and management of the infrastructure lifecycle (Bansal 2021; Borrmann et al. 2009).

BIM and GIS integration enables the use of diverse data sets for sustainability analysis, disaster preparedness, and O&M of an infrastructure. GIS helps in disaster preparedness and to decide where, how, and what types of resources need to be made available to respond. Maintaining information in GIS about contacts of relevant skills, their physical locations, and the experience of disaster response personnel improves disaster preparedness (Amirebrahimi et al. 2016 a,b, Shin et al. 2020). The integration has the potential to make the best possible decisions to support the infrastructure lifecycle at less cost. Neither BIM nor GIS can do this on their own, but the application areas of both have innovative solutions.

Web-based GIS allows stakeholders to achieve a higher level of integration from a variety of web-based sources. This offers an easy way to use available online data and synthesize it with BIM data to facilitate scenario-based planning for decision-making. The integration of BIM and GIS information with time allows a better understanding of the impacts of decisions before, during, and after an infrastructure project is constructed. GIS is a more mature technology used worldwide than BIM. Even though BIM is a relatively new development as compared to GIS, AEC professionals understand the potential benefits that can be realized through the use of BIM. BIM technology is maturing rapidly to deal with the information modeling and interoperability problems in the AEC industry (Ko 2013).

Different information formats are used to store 3D geometry in the AEC industry, however, the two most popular among them are Industry Foundation Class (IFC) for BIM and City Geography Markup Language (CityGML) for GIS. IFC classifies 3D BIM models into four categories according to the detail they contain, using the term *levels of development/detail* (LoD) from LoD_{100} to LoD_{400}. LoD_{100} contains only one solid building in the model; LoD_{400} is a complex model with several components, including interior, doors, windows, walls, balconies, etc. The LoD of surroundings in CityGML for GIS is divided into five different classes (Kolbe et al. 2005). The virtual surroundings become more detailed with increasing LoD both in geometry and thematic differentiation. LoD_0 is a two-and-a-half-dimensional surface model over which an aerial image is draped. LoD_1 has block models of existing facilities without any roof structures or textures over the virtual surface model. LoD_2 has roof structures, textures, and vegetation objects over the virtual surface model. LoD_3 is with roof structures, balconies, bays and projections of existing facilities, and vegetation objects. However, LoD_4 is a walkable virtual model with LoD_3 modeling features and details of interiors like rooms, internal doors, stairs, furniture, etc. BIM and GIS have the same number of levels but have diverse definitions for corresponding LoD. For example, CityGML defines a building model LoD_0 as a footprint of the building, while LOD_{100} in IFC is only one solid building in a model (Zhu et al. 2018, 2019, 2021).

The stakeholders in the different stages need both BIM and GIS together for planning a sustainable lifecycle of an infrastructure project. However, due to several technological barriers, differences in the data types, and file formats supported by BIM and GIS technologies, information from BIM is not easily transferred to GIS or vice-versa (Borrmann and Rank 2009). Such barriers and differences in data types exist because both BIM and GIS have been developed for different purposes by researchers from different backgrounds (Isikdag et al. 2008). As a result, one of these systems does not understand and interpret the information of others with all characteristics. This makes information exchange between BIM and GIS a difficult task (Underwood and Isikdag 2010).

3D model represented with various primitives in BIM, is represented by points, lines, and polygons in GIS. Further, GIS does not support all primitives of BIM, therefore, loss of geometry occurs during the information transfer from BIM to GIS. The transformation of geometric information from BIM to GIS also results in errors like missing elements, textures, and wrong geometries (Underwood and Isikdag 2010). Similarly, during the transformation of surroundings information from GIS to BIM, a loss of semantic information takes place (Isikdag et al. 2008). In addition to this, the local coordinate system used in BIM needs to be manually matched to the geographical coordinate system (GCS) used in GIS. This prevents the seamless integration of BIM and GIS to support lifecycle stages in the AEC industry. However, existing commercial and non-commercial solutions for interoperability between BIM and GIS mainly focus on developing standards to enable seamless data exchange.

BIM tools are used for the calculations of length, area of a region bounded by lines, and the volume of objects. BIM provides a rich set of tools for editing and drafting in comparison to GIS. BIM focuses primarily on 3D modeling, visualization, and analysis as they provide a rich set of tools for digital modeling and analysis; however, infrastructure lifecycle management involves database management, data interpretation, analysis, and simulation. GIS provides several spatial analysis features such as proximity analysis, network analysis, surface area calculation, digital terrain modeling, etc., used in infrastructure lifecycle management in which BIM lacks. BIM provides limited capabilities with respect to non-spatial database management and analysis, while GIS includes a database management system for queries and display of large collections of spatial and non-spatial data on a single platform.

Lastly, BIM is used to model an infrastructure that does not exist yet; therefore, the model is used to demonstrate what is to be constructed. BIM tools are developed to represent the maximum level of detail in terms of geometry and attributes to provide support for the entire lifecycle. On the other hand, GIS is developed to represent site topography, already existing infrastructure, facilities, and utilities. GIS models site topography, already existing infrastructure, facilities, and utilities in the most abstract way in terms of geometry; however, the main focus is on their locations and external dimensions. To represent an infrastructure to be constructed along with its surroundings, a BIM model needs to be placed at its location in GIS, which requires application-level integration between both. Infrastructure development often brings negative impacts to the urban environment in terms of disturbance, traffic jams, noise, dust, and air pollution. The key factor behind these negative impacts is the lack of coordination between stakeholders in strategic planning, which is resolved by establishing interoperability between BIM and GIS (Sebastian et al. 2013).

1.7 Conclusions and Organization of the Book

The use of different software tools to create, analyze, and manage various types of information in the different lifecycle stages of an infrastructure is tedious, time-consuming, and prone to errors. The professionals involved in the different stages have to repetitively develop, organize, interpret, and bridge information which are obtained from using different software tools in different formats. The generation and analysis of the information in a single platform ensures data integrity for planning and management of all lifecycle stages (Bansal and Pal 2008). BIM is used as a single platform for information generation, analysis, and sharing platform during the entire lifecycle of an infrastructure project. However, an infrastructure cannot remain in isolation but is closely related to all other existing facilities and utilities in its surroundings. Hence, the lifecycle of an infrastructure must be planned and managed by keeping in view the site conditions, site constraints, and relationships among existing and the proposed infrastructure. Any new infrastructure addition has an impact on its surroundings; therefore, consideration must be given to the surroundings using GIS. The interaction between existing facilities and proposed infrastructure and its environmental impact are important aspects to be considered for coordinated, integrated, and sustainable infrastructure planning and management. It is like a judgment from a mono-disciplinary to a multi-disciplinary perspective. The use of GIS provides a multi-disciplinary approach for collaborative and sustainable infrastructure lifecycle management. This generates the need for GIS in the AEC industry for infrastructure planning and lifecycle management. The major contribution of the book is threefold. First, a brief introduction with GIS in Chapter 2; second, a detailed discussion on the application areas of GIS, BIM, and integration of BIM and GIS in the six lifecycle stages of infrastructure projects in Chapters 3 to 8; and third, the application areas of GIS in the different lifecycle stages of infrastructure projects that need future research. The book helps in understanding the landscape of work done to use GIS and BIM in the lifecycle stages of infrastructure projects. BIM applications are primarily in vertical infrastructure, like buildings, vertical towers, industrial plants, etc., while GIS applications are closely associated with horizontal infrastructure, like roads, pipelines, railways, canals, dams, utility networks, etc. However, there is no clear-cut boundary line between the two. The book addresses the application areas of both BIM and GIS, thereby encompassing both vertical and horizontal infrastructure domains. However, the primary emphasis of the book is on the application areas of GIS in different lifecycle stages, as numerous books and literature already cover the applications of BIM extensively.

1.8 Exercises

Question 1.1: What is the role of the AEC industry in infrastructure development?

Question 1.2: How do architects, engineers, and construction professionals contribute to infrastructure projects?

Question 1.3: What problems arise in infrastructure lifecycle management when different software tools used by professionals generate different information formats?

Question 1.4: Explain how building information modeling supports information sharing and interoperability throughout the lifecycle of an infrastructure project.

Question 1.5: What are the limitations of building information modeling in the lifecycle management of an infrastructure project that generate the need for GISs in the AEC industry?

Question 1.6: How do BIM and GIS differ in terms of data representation, coordinate systems used, and storage of spatial information?

Question 1.7: What are the benefits of integrating BIM and GIS for sustainable infrastructure lifecycle management?

Question 1.8: What challenges or barriers exist in achieving seamless integration between BIM and GIS software tools?

References

Amirebrahimi, S., Rajabifard, A., Mendis, P., and Ngo, T. (2016a). "A framework for a microscale flood damage assessment and visualization for a building using BIM–GIS integration." *International Journal of Digital Earth*, 9(4), 363–386.

Amirebrahimi, S., Rajabifard, A., Mendis, P., and Ngo, T. (2016b). "A BIM-GIS integration method in support of the assessment and 3D visualisation of flood damage to a building." *Journal of Spatial Science*, 61(2), 317–350.

Bansal, V. K. (2016). "Potential application areas of GIS in pre-construction planning." *Journal of Professional Issues in Engineering Education and Practice*, 142(1), 1–7.

Bansal, V. K. (2017). "Integrated CAD and GIS-based framework to support construction planning: a case study." *Journal of Architecture Engineering*, 23(3), 1–13.

Bansal, V. K. (2021). "Integrated framework of BIM and GIS applications to support building lifecycle: a move toward nD modeling." *ASCE Journal of Architecture Engineering*, 27(4), 1–18.

Bansal, V. K., and Pal, M. (2008). "Generating, evaluating and visualizing construction schedule with Geographic Information Systems." *Journal of Computing in Civil Engineering*, 22(4), 233–242.

Borrmann, A., and Rank, E. (2009). "Topological analysis of 3D building models using a spatial query language." *Advanced Engineering Informatics*, 23, 370–385.

Borrmann, A., Schraufstetter, S., and Rank, E. (2009). "Implementing metric operators of a spatial query language for 3D building models: Octree and B-rep approaches." *Journal of Computing in Civil Engineering*, 23(1), 34–46.

Bowman, D. (1998). "Civil engineering data meets GIS." *Journal of Computing in Civil Engineering*, 12(1), 5–7.

Dore, C., and Murphy, M. (2012). "Integration of historic building information modelling (HBIM) and 3D GIS for recording and managing cultural heritage sites." *18th International Conference on Virtual Systems and Multimedia*, IEEE, 369–376.

Gunes, A. E., and Kovel, J. P. (2000). "Using GIS in emergency management operations." *Journal of Urban Planning and Development*, 126(3), 136–149.

Guo, H. L., Li, H., and Skitmore, M. (2010). "Life-cycle management of construction projects based on virtual prototyping technology." *Journal of Management in Engineering*, 26(1), 41–47.

Isikdag, U., Underwood, J., and Aouad, G. (2008). "An investigation into the applicability of building information models in geospatial environment in support of site selection and fire response management process." *Advanced Engineering Informatics*, 22(4), 504–519.

Jeljeli, M. N., Russell, J. S., Meyer, H. W. G., and Vonderohe, A. P. (1993). "Potential applications of geographic information systems to construction industry." *Journal of Construction Engineering and Management*, 119(1), 72–86.

Jiao, Y., Wang, Y., Zhang, S., Li, Y., Yang, B., and Yuan, L. (2013). "A cloud approach to unified lifecycle data management in architecture, engineering, construction and facilities management: Integrating BIMs and SNS." *Advanced Engineering Informatics*, 27(2), 173–188.

Kang, T. W., and Hong, C. H. (2015). "A study on software architecture for effective BIM/GIS-based facility management data integration." *Automation in Construction*, 54, 25–38.

Kim, H. M., Son, B. H., Kim, Y. C., and Hong, W. H. (2011). "A study of how to improve of building waste management systems using by BIM/GIS." *Journal of Korea Spatial Information Society*, 19(5), 53–62.

Ko, C. H. (2013). "3D-web-GIS RFID location sensing system for construction objects." In *The ScientificWorld Journal*, Hindawi Publishing Corporation, Article ID 217972.

Kolbe, T. H., Gröger, G., and Plümer, L. (2005). "CityGML - interoperable access to 3D city models." Proceeding of *International Symposium on Geoinformation for Disaster Management*, Delft, Netherlands.

Liu, X., Wang, X., Wright, G., Cheng, J. C. P., Li, X., and Liu, R. (2017). "A state-of-the-art review on the integration of building information modeling (BIM) and geographic information system (GIS)." *ISPRS International Journal Geo-Information*, 6(2), 53, 1-21.

Miles, S. B., and Ho, C. L. (1999). "Applications and issues of GIS as tool for civil engineering modelling." *Journal of Computing in Civil Engineering*, 13(3), 144–152.

Ozturk, G. B. (2020). "Interoperability in building information modeling for AECO/FM industry." *Automation in Construction*, 113, 103122.

Sebastian, R., Böhms, M., and van den Helm, P. (2013). "BIM and GIS for low-disturbance construction." In: N. Dawood and M. Kassem (Eds.), *Proceeding of 13*th *International Conference on Construction Applications of Virtual Reality*, 30–31st October 2013, London, UK, 469–479.

Shin, J., Choi, H. S., and You, Y. C. (2020). "BIM/GIS platform-based seismic damage assessment system for building structures." *Journal of the Korean Association for Spatial Structures*, 20(1), 18–22.

Underwood, J., and Isikdag, U. (2010). *Handbook of Research on Building Information Modeling and Construction Informatics: Concepts and Technologies*. Hershey, NY: Information Science Reference.

Zhu, J., Tan, Y., Wang, X., and Wu, P. (2021). "BIM/GIS integration for web GIS-based bridge management." *Annals of GIS*, 27(1), 99–109.

Zhu, J., Wang, X., Chen, M., Wu, P., and Kim, M. J. (2019). "Integration of BIM and GIS: geometry from IFC to shapefile using open-source technology." *Automation in Construction*, 102, 105–119.

Zhu, J., Wright, G., Wang, J., and Wang, X. (2018). "A critical review of the integration of geographic information system and building information modelling at the data level." *ISPRS International Journal of Geo-Information*, 7(2), 66.

Zlatanova, S., and Prosperi, D. (2005). *Large-Scale 3D Data Integration: Challenges and Opportunities*. Boca Raton, FL: CRC Press.

Chapter 2

Geographic Information System

2.1 Learning Objectives

After the completion of this chapter, readers will be able to:

- understand the basics of GIS required in the AEC industry,
- represent different geographical features in GIS, and
- use geospatial analyses in GIS to support different lifecycle stages of an infrastructure.

2.2 Introduction

Infrastructure lifecycle stages require several graphical (drawings, site layouts, charts, blueprints, etc.) and non-graphical (reports, tables, execution schedule, specifications, cost estimates, safety and quality control recommendations, etc.) information. GIS handles both graphical and non-graphical information together on a single platform. This makes it easier to maintain information on infrastructure lifecycle stages in a single central repository. In GIS, graphical information is called geospatial information. These represent the locations, geometrical characteristics, and attributes of an infrastructure and other geographical features on the surface of the Earth. For example, to describe a building in GIS, its location with respect to the geographic coordinate system (GCS), geometry in terms of its physical dimensions, and attributes like name, materials used, and color are represented. The non-geographical information in GIS is called attributes.

Various graphical operations on graphical information and non-graphical operations on non-graphical information in GIS improve and speed up the planning and management of infrastructure lifecycle stages as well as ensure data integrity and accuracy. Reported literature suggests that GIS is supporting the AEC industry in several areas where most of the commercially available BIM tools are lacking. GIS brings geospatial analysis to the real-world spatial reference so that the infrastructure lifecycle can be understood along with its participation in the surrounding landscape, keeping in view the spatial constraints a site offers.

The efforts to use GIS in the AEC industry have not addressed the issues associated with its implementation in the lifecycle stages of an infrastructure. Several

DOI: 10.1201/9781003759348-2

GIS-based tools are suggested in the literature and their applications on infrastructure projects are demonstrated through case studies; their implementations again on real-life projects are not often seen. Future implementations of the developed GIS-based tools in practice still need further consideration. Therefore, professionals in the AEC industry need education and training on the use of GIS in infrastructure projects. The purpose of this chapter is to make the AEC industry professionals aware of the basics of GIS. The research community, software developers, and AEC professionals together may bring GIS as a huge opportunity for the AEC industry.

2.3 Geographic Information System: An Introduction

Geographic information system consists of three words: geographic, information, and system. A system is a group of interacting or interrelated components that work together according to a set of predefined rules or regulations to fulfill the objectives of the system for which it was designed. A system exists as long as its defined objectives are fulfilled. In case the system's objectives are not fulfilled, the system is upgraded or replaced. The most common examples of a system are a computer system, a solar system, a human body system, etc. An information system, as shown in Figure 2.1, is a special class of systems used to achieve the specific objectives of collecting, storing, analyzing, and presenting information in a systematic manner. An information system is operated independently and sometimes is linked with other information systems through standard communication protocols (Lo and Yeung 2002).

Depending upon the class of data that an information system processes, they are classified as *graphical*, *non-graphical*, and *locational* systems. *Graphical* information systems deal with data that is graphical in nature but has no reference to the particular location on the surface of the Earth. For example, the computer-aided drafting (CAD)-based system used to create drawings of a chair, that chair may be placed at any location. *Graphical* information systems are used to process the

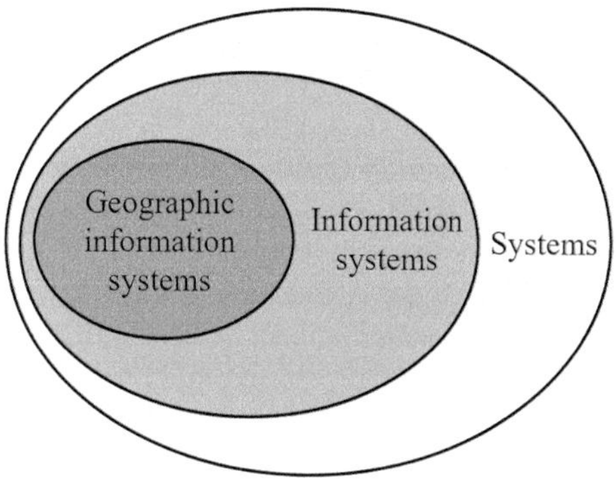

Figure 2.1 Geographic information systems, a special class of information systems

graphical data that has no reference to any position or location on the surface of the Earth. The information systems dealing with graphical data are not considered as a geographic information system. For example, CAD-based systems deal with graphical data but are not GIS. The *non-graphical* information systems are used for dealing with text, numbers, attributes, characteristics, etc. The banking system, accounting system, etc., are examples of *non-graphical* information systems. Finally, the *locational* or *spatial* information systems are used to process the data that have references to locations or positions on the surface of the Earth.

GIS is a system for capturing, storing, checking, integrating, manipulating, analyzing, and displaying data which are spatially referenced to the Earth. GIS is a class of information systems that deals with locations, spatial information, and non-graphical information, as shown in Figure 2.2. GIS deals with locations, spatial information, and attributes together on a single platform. The spatial data relate to the geometry and locations of features on the surface of the Earth. The attribute data, stored in tabular forms, describe characteristics of different features (Chang 2010). Attribute data are also called non-graphical, non-spatial, or non-locational data. GIS is used in different disciplines; accordingly, its capabilities are being expanded. For example, land information systems for land resource management, transportation GIS for transportation applications, and so on.

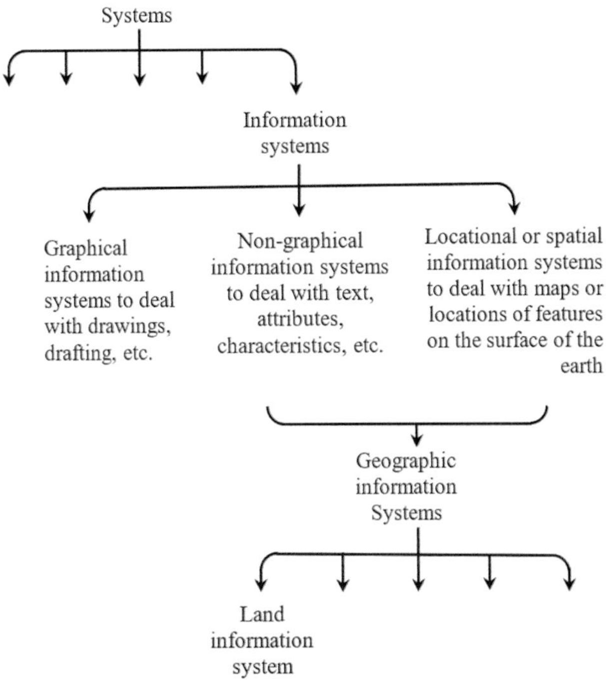

Figure 2.2 Geographic information systems

GIS is normally considered to have a spatially referenced computer database of geographical features on the surface of the Earth. Remote sensing (RS) and the global positioning system (GPS) are the sources of data for GIS. RS is the science of obtaining information about an object, area, or phenomenon through the analysis of data acquired by a device that is not in physical contact with the object, area, or phenomenon under investigation. Even reading a book is an example of RS, the eyes act as sensors that respond to the light reflected from the book pages. GPS is one of the global navigation satellite systems that provides geographic location and time information to a GPS receiver anywhere on or near the Earth's surface where there is an unobstructed line of sight to four or more GPS satellites. GIS provides a medium for storing, manipulating, and analyzing data obtained from RS and GPS.

GIS has four components: hardware, GIS software, human experts, and data to be processed (Clark 2001). Hardware includes computers, input devices, and output devices. The computers are used for data storage, processing, and analysis. The input devices are scanners and digitizers, used for scanning and digitization of spatial data. Other input devices are GPS and mobile devices used for field data collection. The output devices include printers and plotters used for the preparation of reports and hard-copy maps. GIS software includes programs and applications to be executed by the hardware for data management, data analysis, data display, and other tasks. Human experts are the GIS professionals who define the purpose, objectives, procedures, and perform geospatial analyses to present results using GIS software. Data are related to particular processes or phenomena that act as input in GIS. Data used in GIS are available in many formats, like satellite and aircraft images, digital elevation models (DEMs), scanned maps, etc. The data used in GIS may contain information about road infrastructure, communications infrastructure, building footprints, land type and use, water bodies, streams, population densities, etc. Activities of GIS-based geospatial analyses may be classified into spatial and attribute data input, data management, data display, data exploration, data analyses and modeling, and output in the form of maps, tables, graphs, etc. (Lo and Yeung 2002).

2.4 Coordinate Systems in GIS

GIS mainly deals with the locations of various features on the surface of the Earth and their attributes. To define locations of various features on the surface of the Earth, a coordinate system is needed. A coordinate system is a framework for defining the relative locations of various features. The widely used coordinate system to define the locations of various features on the surface of the Earth is the 3D GCS discussed below.

2.4.1 3D Geographic Coordinate System

The Earth looks like a sphere when viewed from space but actually, it is not a sphere. The Earth is wider along the equator than between the two poles. The best

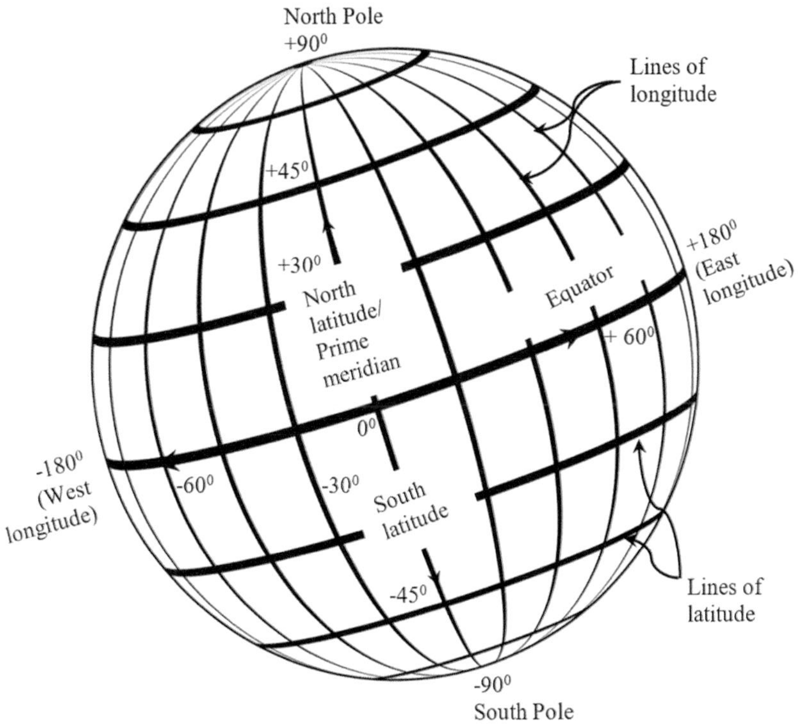

Figure 2.3 3D geographic grid on the surface of the earth, used as the location referencing system

approximation of the shape of the Earth is an oblate spheroid, also called an ellipsoid. The oblate spheroid has two axes: the major axis along the equator and the minor axis along the two poles, as shown in Figure 2.3. Although the Earth's shape is an oblate spheroid but different sizes/shapes of spheroids have been suggested and used in different parts of the world as the best possible model of the Earth's shape to draw the geographic grid system to define relative locations of various features on the surface. The 3D geographic grid system on the Earth's surface is widely used for a location referencing system.

The GCS uses the 3D geographic grid system as the location referencing system for defining locations of different features on the 3D spherical surface of the Earth. The geographic grid system uses longitude and latitude values. Both longitude and latitude readings are angular measures. The longitude values start from 0° to 180° east or west, positive toward the east side and negative toward the west side of the prime meridian. The prime meridian passing through Greenwich, England, is taken as the longitude of 0°. Using the equator as 0° latitude, the latitude values start from 0° to 90° north or south, positive toward the north and negative toward the south of the equator, as shown in Figure 2.4. The latitudes are also called

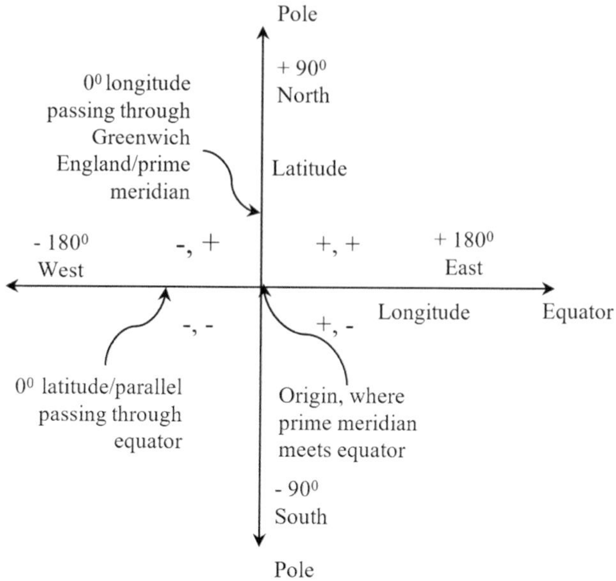

Figure 2.4 Geographic coordinate system

parallels and longitudes are also called meridians. The longitudes are represented along the x-axis and latitudes are represented along the y-axis in 2D, as shown in Figure 2.4. The locations of different features on the Earth's surface based on the 3D geographic grid system are called 3D geographic coordinates and are expressed in longitude and latitude values. For example, the Eiffel Tower, in Paris, France, the most visited tourist attraction in Europe, has a latitude of 48.8584° N and a longitude of 2.2945° E.

2.4.2 *Datums*

The shape of the Earth is an oblate spheroid; different sizes/shapes of oblate spheroids have been suggested and used in different parts of the world as the best possible model of the Earth's shape to draw the geographic grid in 3D. A datum is the model of the shape and size of the Earth used as the reference for the calculation of geographic coordinates in 3D. In other words, a datum is based upon the spheroid adopted for the calculation of the geographic coordinates. Different countries have adopted their own spheroids as the best approximation of the Earth's shape and size. 3D coordinates based upon the spheroid adopted are called geographic coordinates. Clarke in 1866 suggested the spheroid that is used as the standard spheroid for mapping in the United States. North American Datum of 1927 (NAD 27) is based upon the Clarke's spheroid. NAD 27 was later accepted in many parts of the world and used in many GIS software tools.

However, different datums have been suggested and used in the different geographical regions of the world.

With the developments in technology, the shape and size of the Earth were determined with the help of Doppler satellite observations, and the more accurate spheroid, Geodetic Reference System 1980 (GRS 80), was suggested. Based on the GRS 80 spheroid, the North American Datum 1983 (NAD 83) was introduced. Many GIS users have migrated from NAD 27 to NAD 83 or are still migrating because NAD 83 is considered more accurate than NAD 27. In case data are being collected from different sources, it has to be ensured that the data is based upon a common datum in a GIS project. World Geodetic System 1984 (WGS 84) is used by the United States Department of Defense as a global reference system for positioning and navigation. The WGS 84 was established using Doppler satellite observations. It is the datum used in GPS; in other words, all GPS readings are based on the WGS 84 datum. The locational calculations of all GPS receivers are also based on WGS 84. GIS software tools offer a large number of datums to use data from different sources. NAD 83 uses the GRS 80 ellipsoid while WGS 84 uses the WGS 84 ellipsoid; however, the dimensions of these ellipsoids differ slightly. It has to be kept in view that the geographic coordinates of a location change with the datum used.

2.4.3 *Map Projections*

The identification of locations and measurements of distances, angles, and areas on the surface of the Earth is frequently done in the AEC industry. In a 3D GCS, such calculations are difficult; hence, the 3D surface of the Earth is projected on the 2D plane surface. The conversion of the 3D Earth's surface onto a flat 2D plane is called map projection. It converts data sets from GCS to the plane coordinate system. 2D maps are used to represent geographic features on the Earth's surface. The AEC industry professionals generally work with geographic features on 2D maps (flat surfaces) in GIS. In 2D maps, locations of features are based on a plane coordinate system expressed in x- and y-coordinates.

Map projections are generally categorized into two categories based on the preserved property and the projection surface used. The map projection has four classes based on the preserved property: *conformal*, *equivalent*, *equidistant*, and *azimuthal*. A *conformal* map projection preserves shapes, an *equivalent* map projection preserves areas, an *equidistant* map projection maintains consistency of scale along certain lines, and an *azimuthal* map projection retains certain accurate directions. The *conformal* and *equivalent* properties are global properties that are applied to the entire map projection. The *equidistant* and *azimuthal* properties are local and are true only from or to the center of the map projection. The preserved property is important for selecting an appropriate map projection for a project. For example, a population map is based on an *equivalent* map projection, correct area map helps in developing an accurate population density map. However, *equidistant* map projection may be useful for representing accurate distances in the case of missile testing.

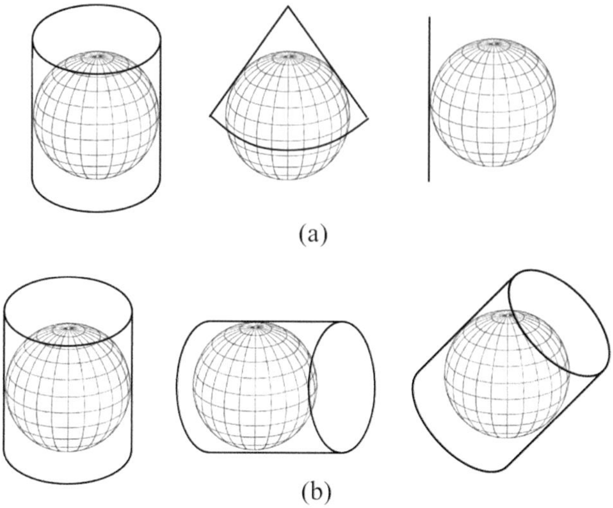

Figure 2.5 Three geometric objects used in map projections: cylindrical, conical, and azimuthal; (a) objects tangent to the globe, (b) cylinder at different positions with respect to the globe

The map projection has three classes based on the projection surface or geometrical objects used for projection: *cylindrical*, *conical*, and *azimuthal*. A cylinder is the projection surface or the developable surface in *cylindrical* projection. A cylinder is placed tangent to a lighted globe and lines of longitude and latitude are traced on it. Other common projection surfaces, as shown in Figure 2.5(a) are a cone and a plane; accordingly, projections are called *conical* and *azimuthal*. A geometric object can have different positions with respect to the reference globe as shown in Figure 2.5 (b). The conversion of a surface from a 3D globe to a plane surface causes distortion in distance, direction, shape, area, etc. no projection preserves all properties. In case some properties are maintained, errors in others are exaggerated. Thousands of map projections are used throughout the world; different map projections preserve different properties and cause different types of distortions. Different map projections are useful for different applications in the AEC industry.

2.4.4 *Projected Coordinate Systems*

Projected coordinate systems and map projections are often used interchangeably. A projected coordinate system is a 2D coordinate system developed on the result of a map projection. The projected coordinate systems are used for locating features in 2D, calculating the coordinates of a location, distance between two locations, area, etc., in 2D. A projected coordinate system in GIS is often divided into different zones to maintain the level of desired accuracy for measurements. Each zone is defined by a different projection center for accurate calculations.

GIS software tools are used to define a projected coordinate system to be used, it involves projecting GCS to projected coordinates or re-projecting projected coordinates from one system to another. A GIS tool typically has many options for datums, ellipsoids, and map projections. Moreover, a projected coordinate system is defined by a GCS and the parameters of the map projection it is based upon. The *Universal Transverse Mercator* (UTM) grid system is a worldwide used projected coordinate system. In case data sets are already projected, re-projection converts data from one system of projected coordinates to another system (Lo and Yeung 2002).

2.5 Modeling of Features in GIS

GIS handles both spatial data (also called graphical or locational data) and attributes (also called characteristics, non-spatial, non-graphical, or non-locational data) together on a single platform. Spatial data, stored in different layers as shown in Figure 2.6, are related to the locations or geometry of different features on the surface of the Earth. On the other hand, attribute data, stored in tabular form, describes characteristics of different features. Corresponding to every layer, there is an attribute table, called the *features attribute table* (FAT) to store attributes. GIS uses a *split file system* or a *single file system* to store spatial and attribute data. GIS stores spatial and attribute data separately in a *split file system* called the *geo-relational* data model. This data model uses an identification number (ID) to link features in a spatial data layer with their attributes in the corresponding *feature attribute table,*

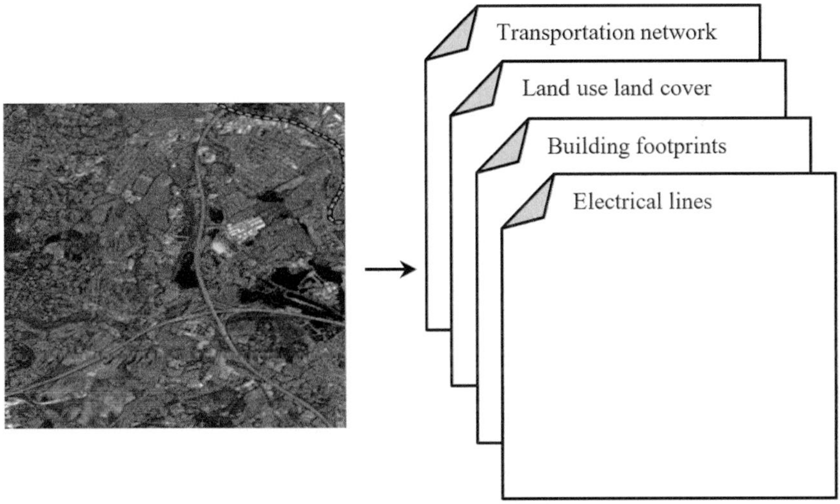

Figure 2.6 Different features on the surface of the Earth are stored in different layers in GIS

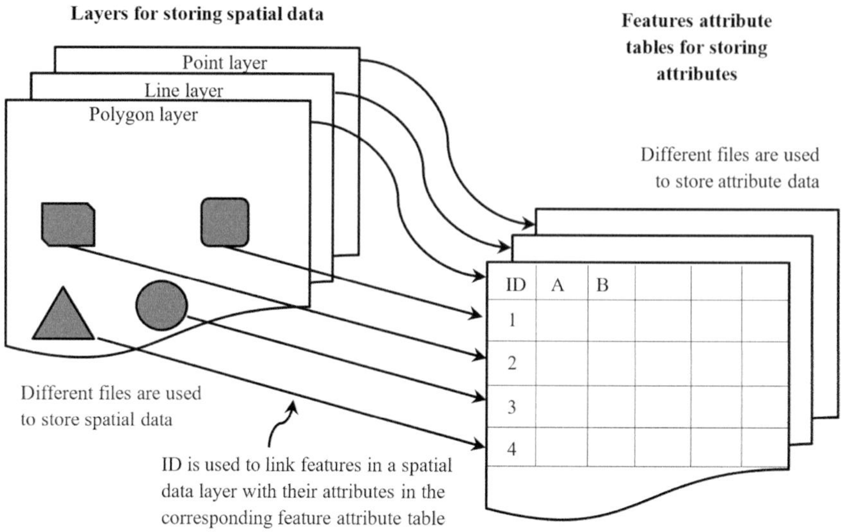

Figure 2.7 Spatial and attribute data in the *split file system*

as shown in Figure 2.7. The features in the spatial data layer are synchronized with their attributes in the corresponding *feature attribute table* so that they are queried, analyzed, and displayed together.

In the *single file system*, GIS uses a single file to store spatial and attribute data. The *single file system* is also called an *object-based data* model, widely used nowadays. The *object-based data* model treats geospatial data as objects. An object is used to represent a spatial feature, such as locations, trees, building footprints, roads, etc., and their coordinate system. It stores geometry and attributes in a single file system. Geometry is stored as a collection of binary data in a special field with data type binary large object. Additional columns are used to store additional attribute data. *An object-based data model* allows a spatial feature to be associated with a set of *properties* and *methods*. A property describes an attribute or characteristic of an object and a method performs a specific action.

2.5.1 *Spatial Data*

In GIS, spatial data are stored in the form of layers. Hence, the locations and geometry of different features on the surface of the Earth are stored in different layers. However, a basic principle in GIS is that all layers to be used together, in a single project in GIS, must be spatially aligned by using the same coordinate system. Different locations and geometries of features on the surface of the Earth are represented with the help of points, lines, or polygons objects in GIS. A layer in GIS either stores points, lines, or polygons, as shown in Figure 2.7. For example, land parcels may be represented by polygons in one layer, whereas centerlines of roads may be represented by lines in the second layer, and locations by using points in the third layer. GIS captures, stores, and manages spatially referenced data as points, lines, or polygons in the vector data model and continuous data in the raster data model. Hence, two fundamental data models for the representation of spatial data in GIS are the vector data model and the raster data model.

2.5.1.1 *Vector Data Model*

The vector data model uses discrete objects to represent features on the Earth's surface. The geometric objects are points, lines, and polygons. A point has zero dimension and has only the property of location. Trees, poles, benchmarks, cities, etc., are represented with point objects. A line is a one-dimensional object and has the property of length, in addition to its location. A line object has a minimum of two end locations and may have additional locations in between to define the shape of a line. The center lines of roads, political boundaries, streams, streets, sewers, water supply lines, etc., are represented with the help of line objects. A polygon object is two-dimensional and has the properties of area and perimeter/length, in addition to locations. A polygon is made up of connected, closed, non-intersecting lines. A polygon may stand alone or share boundaries with other polygons. Areas, land use and land cover, water bodies, counties, forests, soil types, rock types, etc., are represented in GIS with the help of polygon objects. The representation of geographical features on maps depends upon map scale. For example, a city on a map may appear as a point, and the same city may appear as a polygon on another map. A layer in GIS either contains a point, line, or polygon object. Figure 2.8 shows a point layer used to represent locations of trees, a line layer used to represent center lines of roads, and a polygon layer used to represent areas.

Some spatial features are better represented with the help of composites of points, lines, and polygons objects. A triangulated irregular network (TIN) is an example of a composite object used to virtually model hilly terrain. It approximates hilly terrain with a set of non-overlapping triangles. Each triangle in a TIN assumes a constant gradient. The flat areas of the land surface have fewer but larger triangles, whereas areas with higher variability in elevation have denser but smaller triangles. TIN is commonly used for terrain mapping, geospatial analysis, and 3D display. A detailed discussion on TIN has been provided in the chapter titled *Feasibility Stage*.

GIS also provides options for storing both geometry and spatial relationships among different features on the surface of the Earth. The spatial data are called non-topological data if only the geometry of different features is represented. The spatial data are called topological data if geometry, along with spatial relationships among different features, are represented in GIS. Topology ensures data quality and data integrity. The properties and spatial relationships between points, lines, and polygons objects are established in vector data models only. Topology separates GIS from the CAD world. In CAD, the general focus is on the

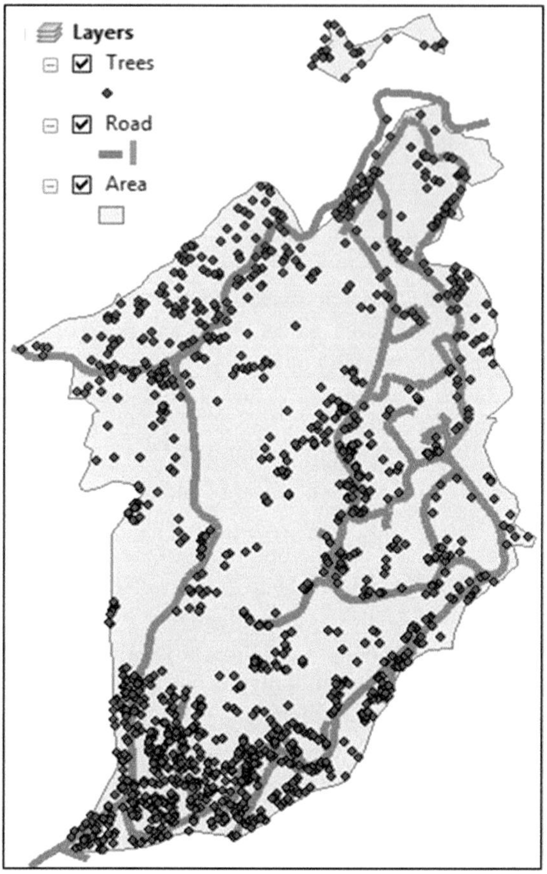

Figure 2.8 A point layer used to represent locations of trees, a line layer used to represent center lines of roads, and a polygon layer used to represent areas

geometrical modeling of features only, not on the spatial relationships. However, non-topological data are displayed more rapidly and easily than topological data. Further, non-topological data are interoperable and used across different GIS software tools. Topology requires additional data files to store spatial relationships. It is in the form of relationship rules, some rules apply to features within a layer, whereas some apply to features of two or more participating layers. For example, topology enables the detection of lines that do not meet, polygons that do not close properly, etc. Topology enhances the quality of geospatial analysis and allows users to perform spatial data queries. A detailed discussion on building topology has been provided in the chapter titled *Feasibility Stage*.

2.5.1.2 *Raster Data Model*

The vector data model uses geometric objects points, lines, and polygons to represent various features on the surface of the Earth. This data model is suitable for discrete features having well-defined locations, geometry, and shapes. However, the vector data model is not suitable for the representation of continuous features or spatial phenomena that vary continuously over the surface of the Earth, such as precipitation, land use and land cover, elevation, soil type, etc. The raster data model is used for continuous features or spatial phenomena that vary continuously over the surface of the Earth.

In the raster data model, spatial data are stored in the form of grid cells or pixels. The Earth's surface or an area is covered by a grid consisting of grid cells, generally of equal-sized square cells, as shown in Figure 2.9. Different cells in a raster are arranged in rows and columns, which are used for data storage and analysis. Figure 2.9 shows points, lines, and polygons objects in vector and raster data models. It indicates how the vector data model is more accurate for representing discrete features. Each cell of a grid stores a value called cell value. The characteristics of an area (represented by cells of a grid) are stored in the form of cell values. Hence, the cell value of each grid cell corresponds to a characteristic of a spatial feature or phenomenon at the cell location. The changes in the cell values of grid cells reflect the variation of a spatial feature or phenomenon over the surface of the Earth. An image is a special case of the raster data model in which a cell value of a cell (attribute) is a reflectance value. The cells of an image are often called pixels. The raster data model uses large amounts of computer memory and takes more processing time than the vector data model.

The cell values of a raster dataset are considered as a matrix used in mathematics; hence, many analyses on raster data are matrix-based. Similar to a matrix, the origin of rows and columns is at the upper-left corner of a raster. The rows of a raster are considered as y-coordinates and columns are considered as x-coordinates. Similar to a matrix, the location of each cell in a raster is defined by its row and column positions. Raster data are available easily and look like an image or photograph. The raster data are easy to produce, process, and have a simple data structure. Raster data are much easier to store, manipulate, aggregate, manage, and analyze than vector data. The common sources of raster data are satellite images (Landsat, SPOT, Digital Globe, Sentinel, Terra Satellite, etc.), digital elevation model, light detection and ranging (LiDAR) data, bi-level scanned files, digital raster graphics, graphic files, digital ortho-photograph, etc. Raster elements used to define a raster are *cell value*, *cell size*, *cell depth*, *raster band*, and *spatial reference*, described below in brief:

Cell value: is an attribute value assigned to each grid cell that corresponds to a characteristic of a phenomenon at the cell location. The cell value applies at the center of a cell, as shown in Figure 2.10. For example, a land use and

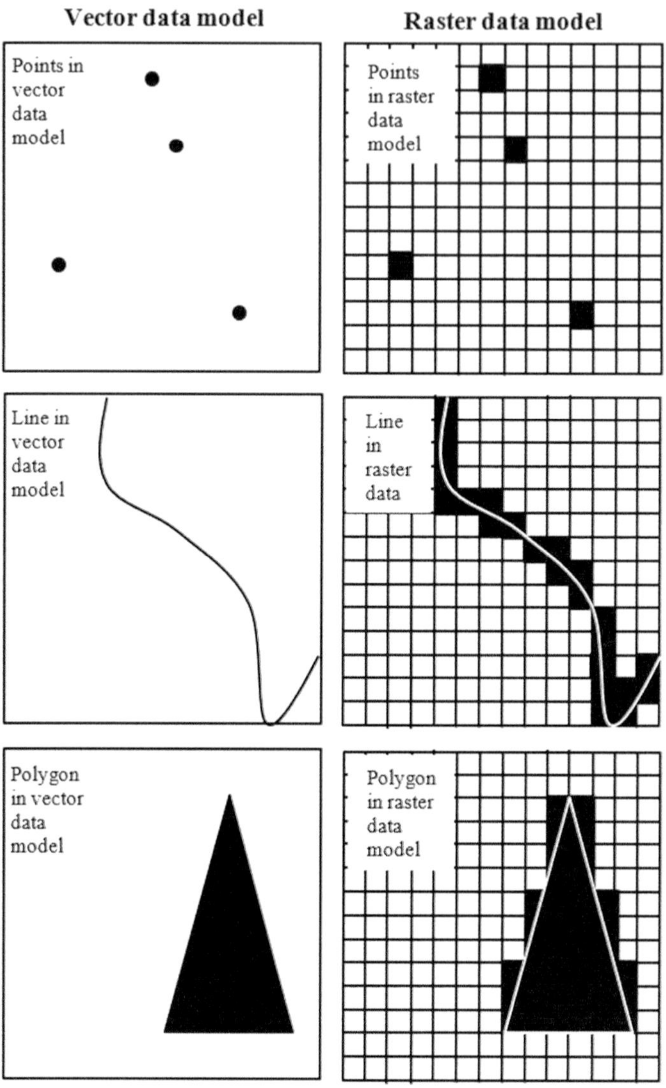

Figure 2.9 Points, line, and polygon objects in vector and raster data models

land cover raster may contain cell value "1" for built-up area, "2" for green land, "3" for water bodies, and so on. The land use and land cover raster is an integer raster because its cell values are integers. On the other hand, the elevation raster contains numeric data such as 800.255, 850.295, and so on, is a floating raster. A floating raster uses more computer memory than an integer raster. Further, an integer raster has a summary attribute table containing cell

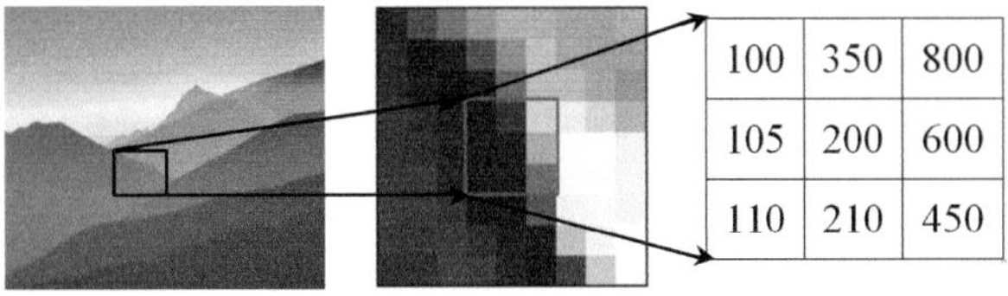

Figure 2.10 Cell values that correspond to characteristics of a spatial phenomenon at the cell locations

values, whereas a floating point raster generally does not because of its large numbers of cell values take up more computer memory. In an integer raster, individual cell values are used to query and display; however, in a floating raster, value ranges are used for query because the chances of finding a specific value are less.

Cell size: is the size of an area represented by a single cell or pixel of a raster. A 100 square meters size of a cell or pixel means each side of a square cell is 10 meters in length. The cell size governs the spatial resolution of a raster. A 10-meter cell-size raster has a higher or finer resolution than a 20-meter cell-size raster. A raster having cells of large size has low resolution, lower spatial accuracy, faster processing, and a small file size. A raster having cells of small size has a high resolution, higher spatial accuracy, slow processing, and a large file size, as shown in Figure 2.11.

Cell depth: is the number of bits used for storing cell values. A bit is the smallest data unit in a computer memory that has a single binary value of either 0 or 1. A byte is a sequence of bits; eight bits are equal to one byte. A higher cell depth indicates that a cell can store a wider range of values. For example, an 8-bit

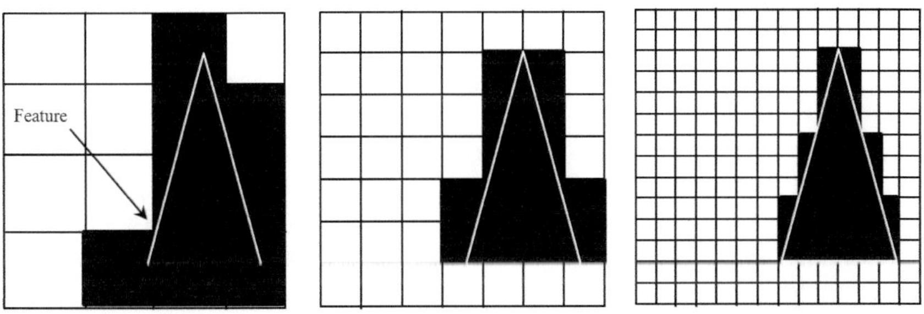

- Large size cells,
- Low resolution,
- Lower spatial accuracy,
- Faster processing, and
- Small file size.

- small size cells,
- High resolution,
- higher spatial accuracy,
- Slow processing, and
- Large file size.

Figure 2.11 Raster images having different cell sizes

raster can store 256 (2^8) possible unique values that range from 0 to 255, while a 16-bit raster can store 65,536 (2^{16}) possible values.

Raster band: a raster is a single band or multiple bands raster. In a single-band raster, each cell has only one cell value. For example, an elevation raster is a single-band raster having one elevation value at each cell location. However, each cell of a multiple-band raster has more than one cell value. For example, an image obtained from a satellite may be a multiple-band raster that has five, seven, or more bands and has five, seven, or more values at each cell location.

Spatial reference: Raster data have the spatial reference information so that they can be aligned spatially with other raster data or vector data sets (in case both are based on the same coordinate system). Spatial reference lets the raster data align or orient with respect to the location or area on the surface of the Earth. The raster data that match a projected coordinate system are called georeferenced raster data.

Raster-based analyses are simple in comparison to vector-based analyses. Neither of the data models is better in every circumstance. The vector data model used to represent discrete features like building footprints has more detail and accuracy than the raster data model. The vector data model stores the information in a more compact format and uses less computer memory than the raster data model (Lo and Yeung 2002). However, the raster-based data model is used to represent the continuous features or spatial phenomena on the surface of the Earth.

2.5.2 Attributes Data

Attribute data describe characteristics of different features in a layer and are maintained in tabular form in GIS. These data use some database management system (DBMS) and are treated in the same way as in other information systems. Corresponding to each layer of spatial data, there is an attribute table called the FAT. An identification number is used to link features of a spatial data layer with their attributes in the corresponding FAT, as shown in Figure 2.7. Each row of the FAT represents a feature in a spatial data layer, or corresponding to a feature in a spatial data layer, there is a row (Chang 2010). The number of rows in a FAT of a layer is equal to the number of features in a layer. Each column of a FAT represents particular characteristics of features in a layer. Attributes may have various data types like strings, integers, floating point numbers, etc. The value at the intersection of a column and a row of a FAT represents a particular characteristic of a feature in the spatial data layer. The query languages and user interfaces in GIS tools permit rapid modification of the values in the FAT. The convenient and quick updating of attribute values are major benefits of GIS over other types of information systems. Figure 2.12 shows spatial and non-spatial data on a GIS platform.

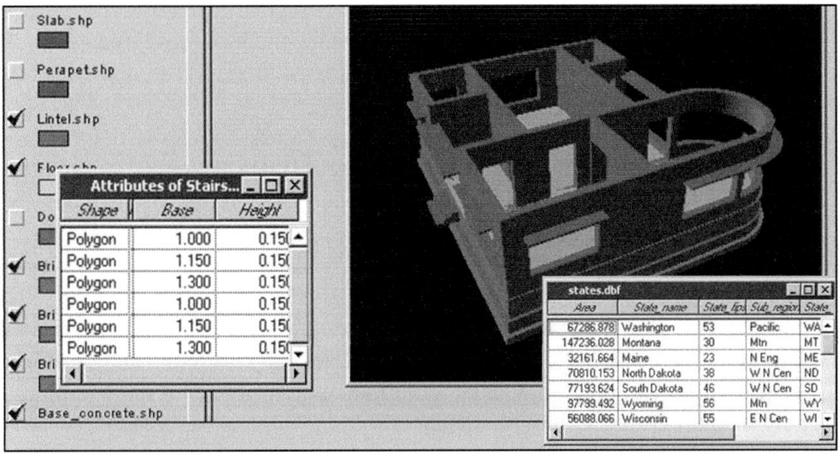

Figure 2.12 Spatial and non-spatial data on a single GIS platform

2.6 Spatial Analysis in GIS

Spatial analysis is the process of analyzing features on the surface of the Earth, phenomena, or processes by examining, assessing, evaluating, and modeling spatial data such as locations, geometry, attributes, and relationships that reveal the geometric and geographic properties of features on the surface, phenomena, or processes. Spatial analysis uses a variety of computational methods, techniques, and algorithms. Spatial analyses are very useful in getting information from the data and making the data easy to understand by segregating different features of interest. The AEC industry has requirements for different types of spatial analyses, which cannot be covered in the present chapter. However, depending on the types of spatial data used in GIS, spatial analyses are categorized into vector and raster-based analyses.

2.6.1 *Vector-Based Analysis*

As discussed in the earlier sections, spatially aligned layers are used to maintain spatial data in the vector data model. Further, the layers to be used together in the vector-based analysis must be based upon the same coordinate system. The vector data model uses points, lines, and polygon layers as inputs in the vector-based analyses. The accuracy of analysis depends upon the accuracy of geometric objects in terms of their locations, geometry, and relationships among them. The vector-based analysis is done on all or selected features in a layer. It is also done on the features of two or more layers together. The widely used vector-based analyses are *proximity analysis*, *buffering*, *overlay analysis*, *pattern analysis*, and *feature manipulation*, discussed below in brief:

Proximity analysis: are the operations used for analyzing the locations of features on the surface of the Earth. Proximity analyses are performed by measuring the distance between features under consideration and other features in their neighborhood. The analysis is generally used to change the locations or geometry of existing features or the creation of new features with respect to distance from the existing features.

Buffering: based on the concept of proximity, buffering creates two areas: one area that is within a specified distance of the selected features and the other area that is beyond. The area that is within the specified distance is the buffer zone. Features for buffering may be points, lines, or polygons. Buffering around points creates circular buffer zones, around lines creates a series of elongated buffer zones around each line segment, and around polygons creates buffer zones that extend outward from the polygon boundaries, as shown in Figure 2.13. The buffers around the geometric objects may be undissolved or dissolved. An object may have more than one buffer zone. For example, a polygon may be buffered with distances of "2", "4", and "6" kilometres, thus, forming multiple rings around, as shown in Figure 2.14. The buffer distance or buffer size may not be constant for all objects in a layer; it can vary according to a field in the FAT.

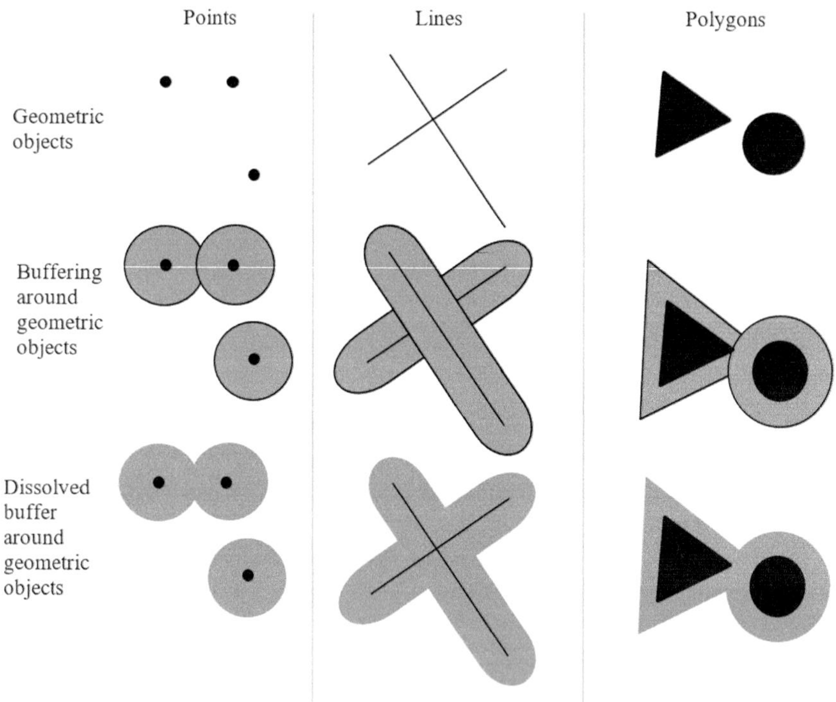

Figure 2.13 Creation of an area of interest around geometric objects: points, lines, and polygons by buffering

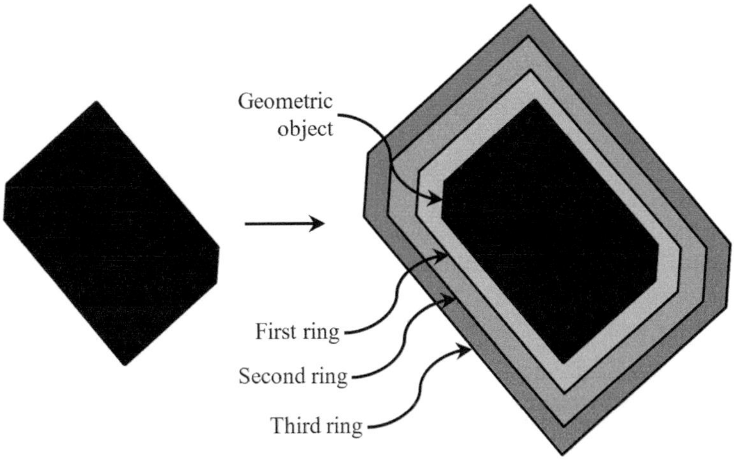

Figure 2.14 Creation of multiple areas of interest around a geometric object by forming multiple rings around

Overlay analysis: involves superimposing two or more layers to produce a new layer that combines the geometric objects and/or attributes of two or more layers. It is a widely used operation in GIS. Varieties of overlay methods are available in the GIS environment for vector-based data, like *polygon-on-point*, *polygon-on-line*, *polygon-on-polygon*, *line-on-line*, *line-in-polygon*, *point-in-polygon*, etc. For example, in *polygon-on-point* analysis, a point layer is overlaid with a polygon layer to determine which points fall within the polygons' boundaries.

Pattern analysis: describes the spatial arrangement of geographical features in a 2D space. Distance measurements are used in spatial statistics to represent distribution patterns. Generally, pattern analysis is the spatial arrangement of point or polygon objects in 2D space. It uses distance measurements as inputs and statistics for representing distribution patterns. Pattern analysis has different levels. For example, at a general level, a pattern analysis reveals that a point distribution pattern is random, dispersed, or clustered. A random pattern is a pattern in which the presence of a point at a location does not encourage or inhibit the occurrence of neighboring points. This spatial randomness separates a random pattern from a dispersed or clustered pattern. The points are spaced at unpredictable distances from each other in a random pattern. In a dispersed pattern, there is measurable space between points. In a clustered pattern, points are grouped in patches or clusters. This is the most common pattern in natural populations. At the local level, pattern analysis can detect if a distribution pattern contains local clusters of high or low values.

Feature manipulation: GIS is used for the manipulation and management of features in one or more feature layers. Feature manipulation is easy to follow graphically, even though various tools available may differ between various GIS packages. Frequently used feature manipulation functions are *appended, clip, dissolve, eliminate, erase, select, split, etc. Append* creates a new layer by piecing together two or more layers. *Clip* creates a new layer containing only those features of the input layer that fall within the area extent of the clip layer. *Dissolve* aggregates features in a feature layer that have the same attribute value or values. *Eliminate* creates a new layer by removing features that meet a user-defined logical expression from the input layer. *Erase* removes those features from the input layer that fall within the area extent of the erase layer. *Select* uses a logical expression to select features from the input layer to create a new output layer. *Split* divides the input layer into two or more layers.

2.6.2 *Raster-Based Analysis*

Raster data structure has fixed cell locations that make it computationally efficient. It facilitates a large variety of analyses in which cell values are used to retrieve useful information. A raster layer is clipped similarly to a vector layer; however, an input raster layer is clipped by a vector polygon layer. The raster clip process results in a single raster that is identical to the input raster but shares the extent of the polygon clip layer. Raster buffers are calculated by developing a grid of distance values equal to the buffer zone from the center of the target cells to the center of the neighboring cells. All the cells of a raster layer are then reclassified. For example, cells having values equal to "1" represent those cells composing the original target cells, cells having values equal to "2" represent cells within the user-defined buffer area, and cells having values equal to '0' represent cells outside of the target and buffer areas. In order to represent the multiple ring buffers, cells of a raster layer are classified by cell values equal to "3", "4", "5", etc., to represent concentric distances around the target cells.

Raster data analyses are performed at the level of individual cells, or groups of cells, or all cells within a raster. Raster analyses may use a single raster or two or more raster layers. Based on the number of cells used, raster analyses are done in different scales of operation: *local*, *neighborhood*, *zonal*, and *global*. *Local* raster operations examine only a single target cell during analysis. *Neighborhood* raster operations examine the relationship of a target cell with proximal surrounding cells. *Zonal* raster operations examine groups of cells that occur within zones or a uniform feature type. *Global* raster operations examine the entire areal extent of the dataset used.

2.6.2.1 Local Operations

Local operations are performed on a single or multiple raster layers in GIS. The local operation uses a mathematical transformation to each individual cell value in a single raster. For example, in an elevation raster in which each cell value represents an elevation in feet, a mathematical conversion is used to convert each cell value in meters. In multiple raster layers, for example, two raster layers contain information on the groundwater depth of an area in the years 2024 and 2025. Just subtracting the corresponding cell values of two layers and storing the difference of cell values in an output raster layer provides the change in groundwater depth between two years. The two main types of local operations used in GIS are re-classification and overlay analysis.

Re-classification: It is usually the first step in raster-based analyses. It involves assigning a new class or range cell value to all cells based on their original cell values. For example, an elevation raster contains a different cell value for nearly every cell within its extent. These cell values are simplified by aggregating each cell value into a few discrete classes. For example, all cells having cell values between 501 and 600 meters elevation are assigned re-classed cell values equal to “1”, between 601 to 700 meters equal to “2”, between 701 and 800 meters equal to “3”, and so on. This facilitates the use of fewer unique values, which are easy to understand and use less storage space.

Overlay analysis: Raster overlay analysis employs powerful *mathematical*, *Boolean*, or *relational* operators to create new output layers. The *mathematical* raster overlay analysis is the most commonly used in GIS. The cell values within the aligned cells of various input raster layers undergo a user-specified mathematical transformation. An output raster layer is created that contains a new cell value for each cell. The *Boolean* raster overlay method uses *Boolean* connectors AND, OR, and XOR to combine the cell values of two or more overlying input raster layers into a single output raster layer. The *relational* raster overlay method utilizes relational operators “<”, “<=”, “=”, “<>”, “>”, and “=>” to evaluate the conditions of the input raster layers. In both the *Boolean* and *relational* overlay methods, cells with cell values that meet the evaluation criteria are coded in the output raster layer with a cell value equal to “1” while those evaluated as false receive a cell value equal to ‘0’. It is important to ensure that all raster layers are spatially aligned, cover identical areas, and have equal resolution. Raster overlay analyses are relatively simple as compared to vector overlay analyses.

2.6.2.2 Neighborhood Operations

Neighborhood operation examines the relationship of an object with similar objects in its surroundings. It examines the target cell value with respect to the cell values of immediate neighbors. The target cell is the cell found in the center of a window. A cell value after the operation is returned based on the combined evaluations, such as minimum, maximum, mean, average, etc., of immediately surrounding the target cell. Neighborhoods are defined by rectangles, circles, wedges, doughnut shapes, etc. The new cell value is placed at the identical location in the output raster layer. Neighborhood operations are commonly used in data simplification.

2.6.2.3 *Zonal Operations*

The zonal operations are used on groups of cells of the same cell value or like features that are in the form of land parcels, political boundaries, water bodies, soil types, etc. These raster versions of groups of cells having the same cell value or polygons are considered as zones in GIS. Zonal rasters are often created by re-classifying an input raster layer into just a few categories of a single raster or two overlaying rasters. Zonal operations measure the geometry of each zone such as area, perimeter, and centroid in a raster. A zonal operation in the case of two input raster layers, one is input raster and the other is zonal raster, the operation produces an output raster that summarizes the cell values in the input raster layer for each zone in the zonal raster layer.

2.6.2.4 *Global Operations*

Global operations are similar to zonal operations in which the extent of the entire raster layer represents a single zone. Typical global operations include determining basic statistical values for a raster as a whole. For example, minimum, maximum, average, range, etc., are calculated over the entire extent of the input raster layer and results an output raster in which every cell contains calculated values.

2.6.3 *Terrain Analysis*

Terrain analysis is the surface analysis in which information related to *slope*, *aspect*, *hillshade*, *viewshed*, *hydrology*, etc., are derived in the form of raster layers. *Slope* maps are used for analyzing and visualizing landform characteristics. Any cell that exhibits a slope is oriented in a direction that is referred to as *aspect*. *Aspect* uses slope information to produce output in the form of a raster layer in which the value of each cell denotes the direction it faces. This is generally in one of the eight ordinal directions: north, south, east, west, northwest, northeast, southwest, and southeast. Sometimes, it is expressed in degrees from 1° (north) to 360° (again back to north). Flat surfaces have no *aspect* and are given a value of "−1". The *slope* and *aspect* information are important because the availability of sunlight varies widely between north and south-facing slopes. This is used in analyzing the stability of slopes to construct an infrastructure. A *hillshade* map represents the illumination of the surface of the Earth from a user-defined sun position. The slope of a hill is relatively bright when facing the sun and dark when facing away. The *hillshade* is calculated using the surface and angle of incoming sunlight light (altitude and azimuth angles) as inputs. *Viewshed* analysis is the technique that uses cell values of a DEM to determine those areas that can be seen from one or more specific locations. The viewing location is either a point or a line placed at any desired elevation. The output of the *viewshed* analysis is a binary raster that classifies all input cells as either "0" or "1", "0" for not visible locations and "1" for visible locations. In the case of two view-point locations are there, the output raster values would be "2" for visible from both viewpoints, "1" for visible from one viewpoint, and '0' for not visible from either viewpoint. The *hydrological* analyses are used to model the flow of water across a surface, derive streams, ridges, flow directions, etc.

2.7 Non-Spatial Analysis

Non-spatial data or attribute data describe characteristics of different features in a layer and are maintained in tabular form. In the vector data model, two types of attribute tables are used: *feature attribute table* and *standalone table*. FAT is linked to the geometry of features in a layer; generally, every vector data layer has FAT. In the case of the *geo-relational* data model, FAT uses the feature ID to link the geometry of features with the corresponding attributes. In the case of the *object-based* data model, the FAT has a field that stores the feature's geometry. FAT also has the default fields that store the geometric attributes of features, such as length for line features and area and perimeter for polygon features. The *standalone table* does not have direct access to the geometry of features but has a field to link it to FAT whenever required. A DBMS is used to manage FAT and *standalone* tables in GIS.

The manipulation of attribute data in GIS includes deleting or adding fields from FAT and *standalone* tables. Deleting a field is straightforward that requires specifying an attribute table and the field to be deleted. The addition of a field is required for the storage of additional characteristics, classification, or computation of additional attribute data. The new field is designed to receive the result of classifications or computations. New attribute data are also created from existing data using mathematical expressions. Operationally, it involves defining a new field and computing its values from the values of existing fields in an attribute table.

2.8 Conclusions

Many of us in our routine lives are using geospatial technology. For example, to find the location of a nearby park, an online search is done. Search provides the name of the nearby existing parks, their locations on a map, their distances from the current location, and attributes. Once the park is to be visited is decided it provides the shortest route for driving. To reach the decided location, a navigation system is used to get the driving directions and to record places visited photographs are geo-tagged. All of these activities involve the use of geospatial technologies. Geospatial technologies are being used, but they do not require a GIS professional. However, to use GIS in the AEC industry, professionals must be familiar with the GIS technology as well as the basic concepts that drive it. The purpose of the chapter is to provide GIS concepts for AEC professionals. A GIS is a computer system for storing, managing, analyzing, and displaying geospatial data. Geospatial technology centers on GIS and uses GIS to integrate data from RS, GPS, cartography, and surveying to produce useful geographic information for the AEC industry.

2.9 Exercises

Question 2.1: What types of information are used in the infrastructure lifecycle stages and how does GIS handle them?

Question 2.2: Why is GIS considered an advantage over commercially available BIM in the AEC industry?

Question 2.3: Why is a coordinate system necessary in GIS, and what are the differences between geographic and projected coordinate systems?

Question 2.4: Describe the concept of datum in GIS and explain how different datums affect geographic coordinates.

Question 2.5: What are map projections? Why are they important? Provide an example of a suitable map projection for a specific application.

Question 2.6: Differentiate between a split file system and an object-based data model used for storing spatial and attribute data in GIS.

Question 2.7: What are the three geometric objects used in the vector data model to represent Earth's surface features? Give two examples of features represented as point, line, and polygon objects in GIS.

Question 2.8: What is a TIN, and what are its common uses in GIS?

Question 2.9: Differentiate between topological and non-topological data in GIS. Why is topology considered important in GIS compared to CAD-based systems?

Question 2.10: Why vector data model is more suitable for discrete features on the Earth's surface, while the raster data model is more suitable for continuous features in GIS?

Question 2.11: In raster data models, how are spatial data stored, and what does a cell value represent?

Question 2.12: Define cell size in the raster data model and explain how it affects resolution and file size.

Question 2.13: What is meant by "cell depth" in a raster data model? Provide an example comparing 8-bit and 16-bit raster data.

Question 2.14: Differentiate between a single-band raster and a multi-band raster with examples.

Question 2.15: What are the advantages and disadvantages of the vector and raster data models in terms of accuracy, memory usage, and processing speed?

References

Chang, K. T. (2010), *Introduction to Geographic Information Systems*, 4th ed. New Delhi, India: Tata McGraw-Hill.

Clark, K. (2001). *Getting Started With Geographic Information Systems*. Upper Saddle River, New Jersey: Prentice Hall.

Lo, C. P., and Yeung, A. K. W. (2002). *Concept and Techniques of Geographic Information System*. New Delhi: Prentice-Hall of India.

Chapter 3

The Feasibility Stage

3.1 Learning Objectives

After the completion of this chapter, readers will be able to:

- understand the use of GIS for database management,
- model the surroundings of an infrastructure and physical constraints on a site,
- understand the application areas of GIS in site selection, site access planning, extending services to a construction site, and
- understand the application areas of BIM and the integration of BIM and GIS in the feasibility stage.

3.2 Introduction

The feasibility stage involves a comprehensive analysis of an infrastructure project in order to evaluate its technical viability, economic feasibility, and mainly the chances of its success or failure. Infrastructure projects are cost-intensive, in the feasibility stage, a detailed assessment to analyze and finally decide whether to proceed or not to proceed with the proposed infrastructure project is done. It is ensured before the project execution that the project will not waste important resources like time, cost, and energy. Large-scale infrastructure projects have various other aspects to be taken into consideration before execution like environmental impact assessment (EIA), obtaining permissions from different authorities, regulatory approval, financial approval, etc., before going ahead with the final decision on the investment. GIS helps in the feasibility stage in determining whether or not the project ideas have the potential for success based on market analysis, site constraints required to be fulfilled in order to complete a project, and if another location is better for a proposed project what are the chances of success and potential return on the investment. The ability of GIS to process the information quickly allows the AEC industry to improve decision-making in the feasibility stage. It promotes better organizational integration and knowledge management. GIS is not only used for the technical evaluation of an infrastructure project but also in the study of the sociological and economic feasibility of different proposals (Jeljeli

DOI: 10.1201/9781003759348-3

et al. 1993). This chapter, in the following sections, covers the application areas of GIS, building information modeling (BIM), and the integration of BIM and GIS in the feasibility stage of infrastructure projects.

3.3 GIS in the Feasibility Stage

GIS supports infrastructure projects' feasibility stage by managing required spatial (site maps, topography, locations of existing facilities/utilities, subsurface profiles, etc.) and non-spatial data. Through the modeling and geospatial analyses, GIS improves decision-making, accuracy, and efficiency in the following areas:

3.3.1 Database Management

The different types of information used in the feasibility stage of an infrastructure project are broadly categorized into graphical and non-graphical. The graphical information is also known as spatial or locational information in the GIS community. The non-graphical information is also known as non-spatial information, non-locational information, attributes, or characteristics. Infrastructure projects contain both types of information, GIS has the capability of dealing with both types of information. The following subsections describe the capability of GIS to handle both types of information in infrastructure projects:

3.3.1.1 Spatial Data Management

GIS is also called a spatial data management system due to its capabilities to acquire, store, analyze, and display spatial data. The spatial information in an infrastructure project may include:

- topography, geology, contour map, hydrology of project sites, etc.
- layouts of service networks: electricity, water supply, sewer disposal, gas supply line,
- location map of various existing services and utilities, and
- geometry of an infrastructure project under consideration, etc.

A lot of information is available online and can directly be used as input in GIS for strengthening decisions in the feasibility stage. Examples of online available graphical information are: satellite imagery, topography, digital elevation models, geological maps developed by local government bodies, etc. Information is also obtained from remote sensing, aerial photography, and field surveys using instruments like total station and global positioning system. Varieties of graphical information are created in GIS itself by digitizing, scanning maps and drawing, etc. GIS maintains graphical information by using vector and raster data models as discussed in detail in the previous chapter.

In the feasibility stage engineers, architects, or planners have to repetitively collect, reorganize, interpret, and analyze graphical information collected from various resources. This is a time-consuming process that is quite tedious and prone to errors (Adams 1994). GIS capability to maintain large graphical databases in the digital form is utilized to maintain graphical information required in the feasibility stage. The availability of information in the digital platform provides the AEC industry with a mechanism for rapid retrieval and doing required manipulation (Oloufa et al. 1992).

The construction site's underground conditions are investigated by making boreholes at different locations. GIS is used to develop a database to store descriptive soil data that is linked to the corresponding locations of boreholes. The information about the surface and subsurface conditions of a site after analyzing the borehole data are maintained in GIS. GIS is used to develop the database for deciding the type of foundation, foundation analysis, design, construction planning, and design-construction integration in infrastructure projects (Oloufa et al. 1994). The surface and sub-surface conditions of a site which are maintained in GIS influence construction methods and the choice of equipment to be used that in turn affect project cost. GIS facilitates the development of a graphical user interface to input, query, and take output from the developed databases.

The AEC industry benefits greatly from 3D geotechnical subsurface information databases. Maintaining sub-surface information digitally in a relational database, rather than as images not only allows the distribution of bore logs data over the internet but also provides an opportunity to explore that and do planning online. The web-based GIS databases allow for performing online spatial queries, generation of bore logs, generating vertical cross-sections of soil profiles along any arbitrary line, and performing various geotechnical analyses (Kunapo et al. 2005). It provides easy-to-use functions to locate boreholes using search and geotechnical spatial queries to get valuable information that is of immense value in the feasibility stage.

The integration of a spatial database with the project management functions provides a powerful and effective project management system. The integrated project data is visualized instantly by color-coding the spatial display based on the non-spatial attributes in the GIS environment (Sun and Hasell 2002).

3.3.1.2 *Attributes Data Management*

Non-graphical data of infrastructure projects are divided into four categories: geographically referenced data, non-spatial attribute data, geographic indices, and spatial relationships. Geographically referenced data describe information about a specific location in text, such as locations of availability of raw materials, shops, etc. Non-spatial attribute data describe attributes of features in a layer, such as characteristics of raw materials available at a location, different types of materials available in shops, etc. Geographic indices describe the details of geographic locations, such as addresses of raw materials locations, addresses of shops, etc. Spatial relations describe the proximity of features to one another. The non-graphical data in GIS is generally stored in tabular form. GIS software tools generally have two types of tables; *features attribute tables* and *standalone tables*, as discussed in Chapter 2. The *standalone tables* do not have corresponding layers of graphical features.

Different types of labor are employed in infrastructure construction projects for various tasks. Various labor types and their daily wages are maintained in *standalone tables*. A planner adds or deletes labor types from *standalone tables* depending on the specific project requirement. The planner also modifies daily wages from time to time and for different project locations. Infrastructure projects are highly equipment-intensive. Equipment has virtually replaced large groups of labor involved with earlier methods of construction. The size and capacity of equipment being used determine the time duration, and cost of a construction project. Equipment information is also maintained in a *standalone table* in GIS. Equipment information is updated at any point of time by a planner during the planning or construction stage. *Standalone tables* are used to maintain the information about the costs of different types of construction materials used. Materials rates usually exclude the cost of carriage to a site. The carriage, loading, and unloading costs are added to the materials costs to arrive at the total costs of materials on a site (Bansal and Pal 2006).

Each infrastructure project is different from the other with respect to unique project specifications. The project-specific information is used as initial input for the computation of work quantities and costs of various tasks involved in a project. The *standalone tables* are used to store attributes like: execution schedule, quality control and safety recommendations, list of tasks to be performed, overhead costs, contractor's profit, etc. The tasks to be performed in a project are derived from the design drawings, proposed construction elements, and construction methods. Each task includes a set of activities which are carried out in sequence to complete the task.

3.3.1.3 Spatial and Attribute Data Management

Spatial and non-spatial information are stored together in the GIS environment. GIS provides a data model to store, retrieve, manipulate, and analyze both spatial as well as non-spatial information together. It has features like the definition of attribute types, query processing, and multiuser operation processing. Road construction involves large volumes of information on design, construction methods, work quantities, unit costs, production rates, and site conditions. It is important to acquire, manage, and process necessary information efficiently to produce a rigorous construction plan. GIS is used for managing and integrating spatial and non-spatial information required for road construction planning. GIS-based database management systems are used in road construction planning with their interactive space scheduling and operation-level planning functions, which were supported by the integration of various data required for planning (Seo and Kang 2006). GIS enables planners to review the construction plans intuitively and immediately by querying graphical objects and corresponding attributes. Spatial and non-spatial databases in GIS help in the calculation of areas, volumes, and perimeters as well as cut and fill volumes. GIS is being explored for the AEC industry in the development of a single database that can be used in the feasibility stage.

3.3.2 *Modeling Surroundings of an Infrastructure*

The visualization and knowledge of a terrain play a significant role in site selection, designing, construction planning, as well as in work execution. This section provides an overview of how GIS is used in the virtual modeling of terrain and performs different types of analyses.

3.3.2.1 *Surface Modeling*

The virtual 3D surface model of an area is the digital representation of its natural topography. It gives a perspective of the difference in heights between valleys and ridges in an area. To generate the surface model of an area, latitude, longitude, and elevation values on a few locations on the surface of the area are taken with the help of a total station. The accuracy of a virtual surface model depends on the number of values taken at different locations in an area. The input used to develop a surface model in GIS is a table that has a minimum of three fields: latitude, longitude, and elevation. The elevation value varies for every location on the surface of an area. To determine the elevations of other points on the same surface, nearby measured elevation values are interpolated. Another alternative to develop the surface model in GIS is the use of a contour map of an area as input. In GIS, two types of virtual surface models are used to represent a surface: *triangulated irregular network* and *elevation raster*.

To represent a surface virtually, TIN is a more accurate model than an elevation raster. TIN is used for high-precision surface modeling of small areas for engineering calculations. It is a vector-based representation of the surface made up of irregularly distributed nodes with their coordinates (latitude, longitude, and elevation) arranged in a network of non-overlapping triangles. Nodes of a TIN remain on the entered locations taken with the help of a total station to preserve their locational precision. An edge is formed by connecting two nodes, and has an elevation value at each end, which helps in calculating the slope along the edge. Each triangular facet of a TIN describes the behavior of a small part of the virtual surface. The coordinates at three nodes of a triangle are used to derive information about slope, aspect, surface area, and 3D surface length. TIN facilitates in deriving information about areas, volumes, surface profiles, visibility analyses, etc., in the feasibility stage.

In the first step, data from a total station are transformed into GIS with the help of a compatible software tool. The tool used to transfer data from the total station depends on the total station itself and GIS software tools being used. GPS is also used as an input data source for creating a surface model. The latitudes and longitudes values of different locations in GIS are converted to a two-dimensional (2D) layer. The common format used to store a layer in GIS is a *shapefile*. The *shapefile* is a simple non-topological format for storing geometric locations and their attributes. In the second step, the latitude and longitude values of each location are assigned its elevation value. A point in a layer represents a location on which latitude, longitude, and elevation are taken with the help of the total station. Points in a 2D layer are converted into 3D by assigning their elevation values from the Feature

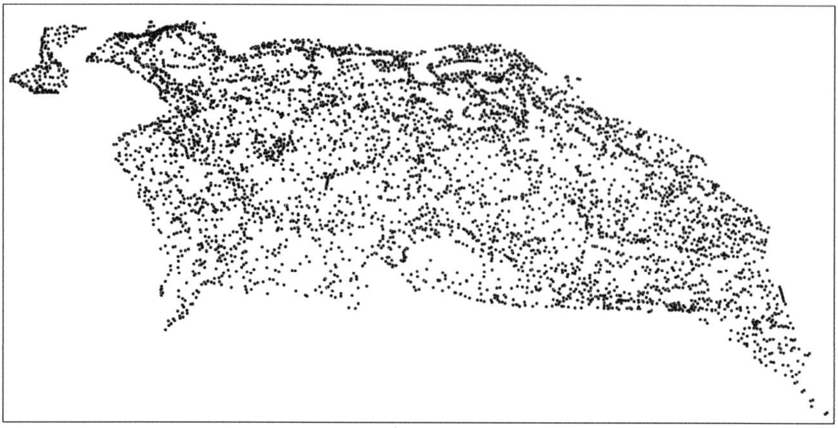

Figure 3.1 A GIS layer representing locations of various points in 3D

Attribute Table (FAT). A layer representing the locations of various points in 3D is shown in Figure 3.1. In the third step, an elevation raster or Triangulated Irregular Network (TIN) is generated by taking 3D points as input. The elevation raster is a rectangular array of cells in which each cell represents a square area on the surface of the earth and has an elevation value that is static for the entire cell area, as discussed in Chapter 2. The locational precision of a surface model is related to the size of a cell in an elevation raster.

Figure 3.2 shows a 3D view of the developed TIN. Figure 3.3 represents the same digital surface in 2D. Figure 3.3(a) is an elevation raster constructed in a grid format with uniformly spaced cells, and Figure 3.3(b) is a TIN constructed by triangulating 3D points (Bansal 2014). Both the elevation raster and TIN allow deriving

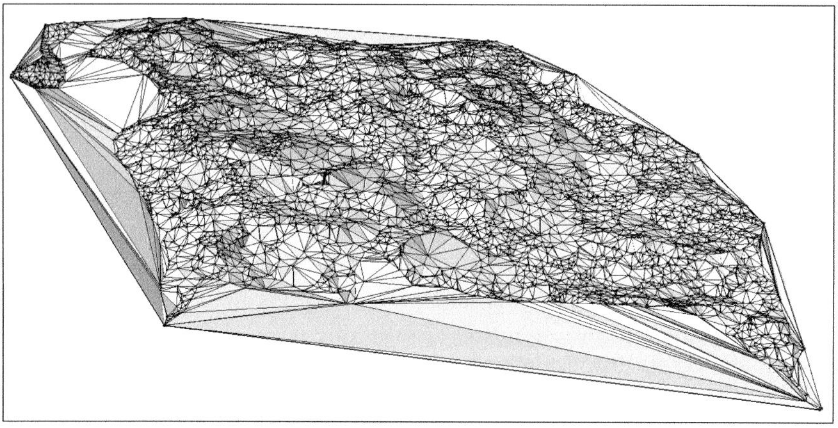

Figure 3.2 A 3D view of a triangulated irregular network

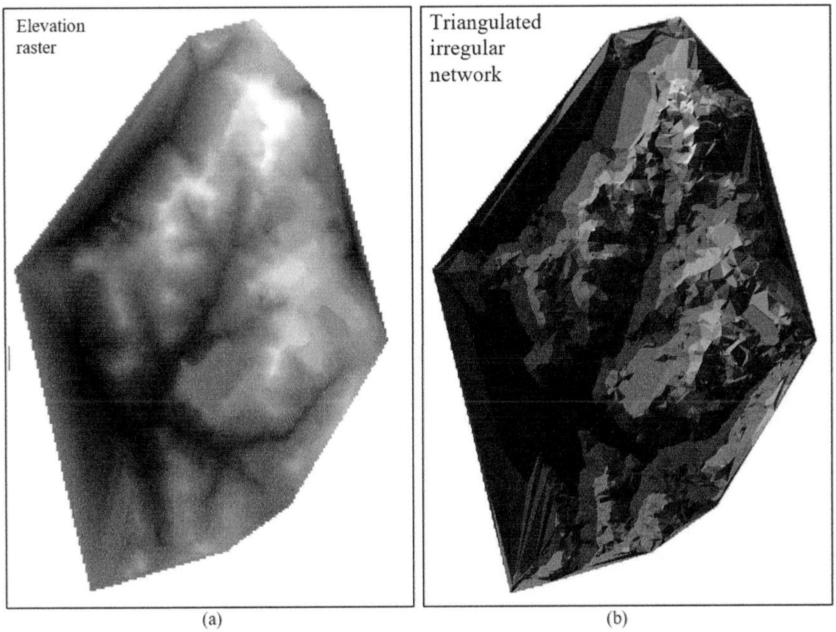

Figure 3.3 2D digital representation of a surface: (a) a raster constructed in a grid format with uniformly spaced cells and (b) a TIN constructed by triangulating 3D points

the slope and aspect of a site. The rate of change of elevation is called slope while the directional measure of slope is called aspect. Slope and aspect information are used in slope stability analysis also. To give a more realistic appearance to the virtual surface, a geo-referenced aerial photograph of the same area is draped over it using suitable ground control points.

3.3.2.2 Modeling of Existing Facilities and Utilities

In order to virtually model existing facilities and utilities in an area, an aerial image of an area is geo-referenced and footprints of existing facilities and utilities are digitized from the image. Alternatively, CAD-based drawings representing footprints of existing facilities and utilities are imported into GIS. CAD drawings are aligned in GIS by using suitable ground control points. The digitized footprints are stored in a layer in GIS. The elevation values on the footprints are obtained from the developed surface model. GIS does not have user-friendly 3D modeling capabilities; therefore, 2D footprints digitized in GIS are exported to a suitable 3D modeling software tool along with their elevation values and geo-referencing system for 3D modeling.

All existing facilities and utilities are externally modeled on their respective footprints in a 3D modeling software tool. The selection of the software tool depends upon the interoperability between the 3D modeling tool and the GIS tool being used. The low-pixel digital camera is used to acquire images of all existing buildings at least in four cardinal directions. These images are pasted on the respective faces of a 3D model to give it a more realistic appearance. Once modeling is complete, 3D models of existing facilities and utilities are imported back to GIS. Existing facilities or utilities are modeled with a minimum level of internal details in terms of their geometric representation to keep file size small and processing fast (Bansal 2021).

Information in digital form is important during the lifecycle of an infrastructure. However, there is a challenge in the information exchange between 3D infrastructure modeling software and GIS, resulting in the requirement of remodeling of the digital information in GIS for many geospatial analyses. Knoth et al. (2019) suggested three major sources for infrastructure modeling in 3D and transformation processes to enable the integration of developed models into the geographical environment in the form of one transition model.

3.3.2.3 *Virtual Modeling of Surroundings*

The virtual model of the surroundings of an infrastructure is developed by placing 3D models of existing facilities and utilities over the 3D surface model. It facilitates in viewing an area from any direction, height, and inclination. The use of a 3D virtual model of the surroundings of an infrastructure is more effective in different types of planning rather than the use of contour maps, locational maps, 2D drawings, or 3D wooden/plastic models when synchronization with nature and environmental impact is of great concern. It gives a better insight into the topography and surroundings of an infrastructure of interest to reveal spatial relationships. Figure 3.4 shows a perspective view of the virtual surroundings, a planner may walk through and have a virtual visit of the site for quick reference.

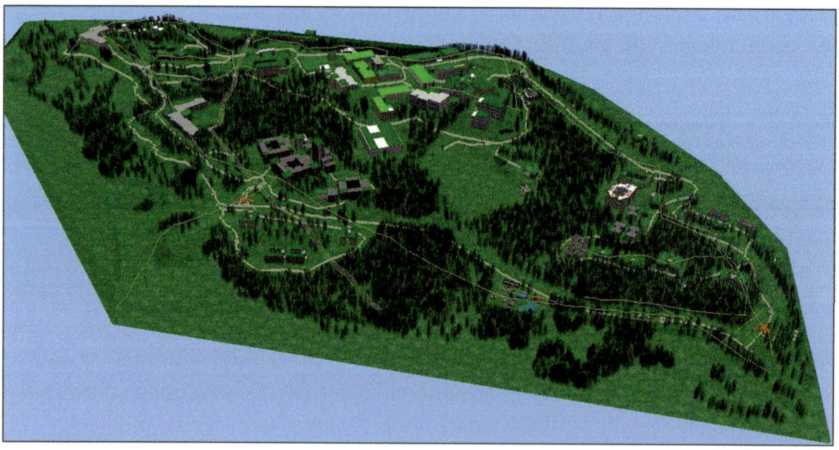

Figure 3.4 A perspective view of virtual surroundings

Figure 3.5 A perspective view of virtual surroundings with LoD_3

The levels of detail (LoD) are classified into five classes (Kolbe et al. 2005). The objects become more detailed with increasing LoD both in terms of geometry and thematic differentiation. LoD_0 is a two-and-a-half-dimensional digital terrain surface model over which an aerial image is draped. LoD_1 has block models of buildings without roofs or texture details over a LoD_0. LoD_2 has roofs, textures, and vegetation objects. The roof structures, balconies, bays, and projections of existing facilities and vegetation objects having LoD_3 are shown in Figure 3.5. LoD_4 is a walk-able virtual model with LoD_3 modeling features and details of interiors like rooms, internal doors, stairs, furniture, etc. The view of a building in its surroundings having LoD_4 is shown in Figure 3.6. It gives a better insight into the locational aspects of a site to reveal spatial relationships among nearby existing utilities/services and infrastructure under consideration.

GIS provides geo-visualization and animations for the exploration of virtual surroundings and their components with the help of object-oriented techniques. GIS helps in the development of visualization of virtual surroundings and their components in 2D, 3D, and time-varying animations (Döllner and Hinrichs 2000). The advantages of visualization of geographical conditions and their impacts on different stages of an infrastructure project are explored by using time-lapse simulation (Sonmez and Uysal 2008).

3.3.2.4 Geospatial Analyses on Surroundings

The virtual model of surroundings helps in performing many geospatial analyses in the feasibility stage. The location-based selection helps in selecting a

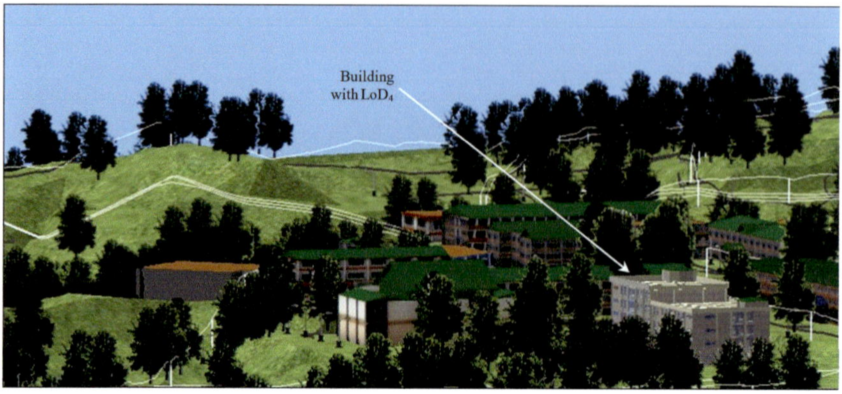

Figure 3.6 A building with LoD_4 in its virtual surroundings

facility or utility based on its location relative to other existing infrastructure. The selection based on area, highlights all facilities or components of facilities that occur within a specified distance. In an area-based selection, buffers of a specified distance are created around the source features. As a result, selection highlights all components or facilities in the buffer zones. The virtual surface model is used to calculate projected 2D area, 3D surface area, and volume. The surface area is measured along the slope of a surface by considering variations in elevation values. The difference between surface area and 2D projected area gives an indication of the roughness or slope of a surface. A large difference means a rough or a steep sloped surface and a small difference means a smooth or a gentle sloped surface. The volume occupied between a given surface and a reference plane at a particular elevation is also calculated. This helps in the computation of excavation work between a plane and a surface. The volume of water stored on the upstream side of a check dam is also calculated using a surface model as an input.

The virtual surface is used to develop a profile along a line that provides changes in elevation along a line on a surface. This makes it possible to assess the feasibility of constructing a new road, railway, or canal along a line. Vertical profiling is used to visualize topography along a proposed route. The route is drawn on a virtual surface model to create a vertical profile. The profile shows changes in elevation along the route as per the topography. Profiling helps in selecting an optimum route for highway alignment, earthwork haulage, and transporting equipment to reach at work sites. The number and magnitude of ups and downs in a profile graph decide the selection of a route among various alternative routes based on the grade resistance offered by each one. A profile along a line AB on the virtual surface model is shown in Figure 3.7.

The virtual model of surroundings helps in performing visibility analysis; it generates a line of sight between two points lying on a surface showing the visible

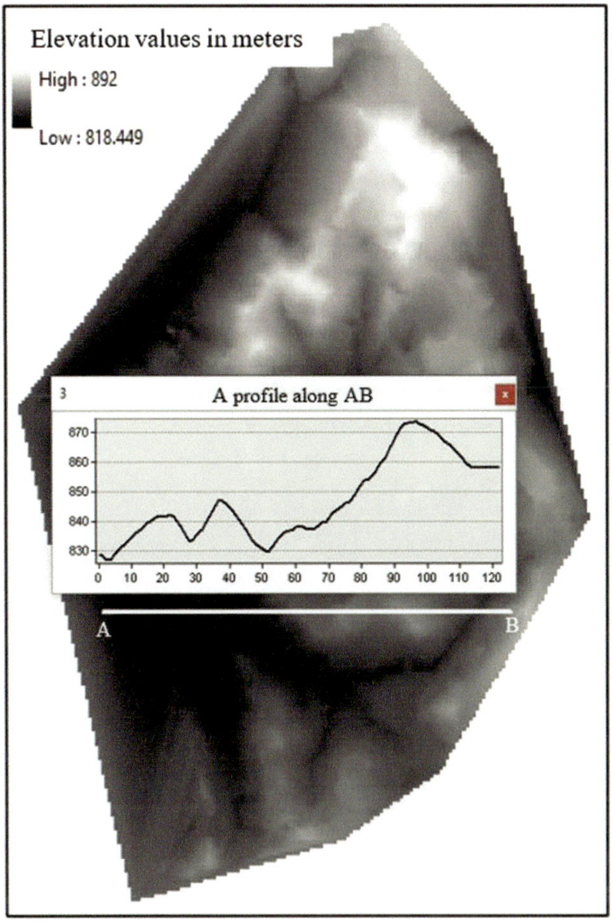

Figure 3.7 A profile along the line AB on the virtual surface

and obstructed points. The points on a line of sight where a surface is visible or obstructed are shown with two different colors. Figure 3.8 shows visibility analysis along a line showing obstructed points in gray color and visible in white color. It is possible to generate a contour (a line connecting points of equal elevation) passing a specified location as shown in Figure 3.9. The developments in hill areas require an understanding of the natural flow of water and how any change in the topography because of infrastructure development affects water flow patterns. A planner or an architect easily understands the inflow and outflow of water in an area using its virtual surface model. The water follows the steepest path downhill until it reaches the perimeter of a surface model or a pit as shown in Figure 3.10. The logistics planning is mostly ignored in the feasibility stage. The information for logistics

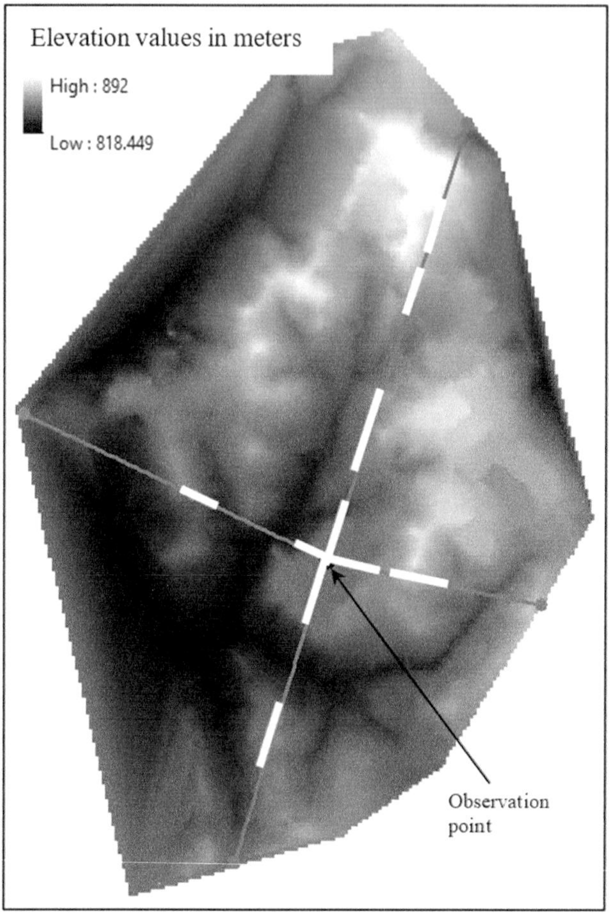

Figure 3.8 Visibility analysis from an observation point: obstructed sights are shown by thin lines in gray color and thin lines in white color show visible sights

planning is stored in various forms, thus making its in-time retrieval difficult. The surface model developed in GIS facilitates the calculation of distances between different locations, considering the path distance or travel time as road impedance (Varghese and O'Connor 1995).

3.3.3 Modeling Surface and Sub-Surface Conditions

Site investigation is an important step for checking feasibility and in the construction planning of new infrastructure projects. The database management capabilities of GIS are used to develop sub-surface profiles from well-log data

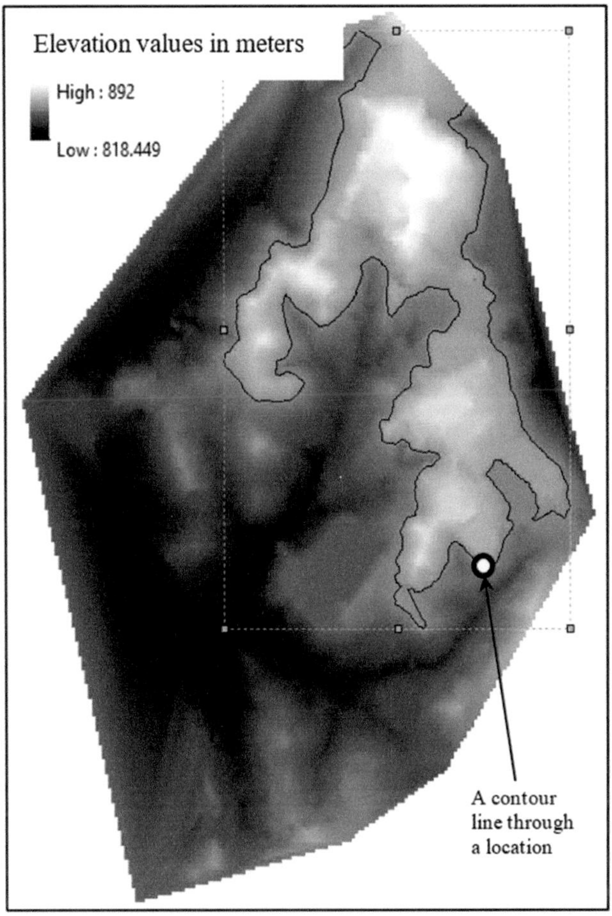

Figure 3.9 A contour line passing through a specific location

(Camp and Brown 1993; Parsons and Frost 2000). The well-log database is constructed from a series of borehole data. A 3D sub-surface environment or 3D sub-surface profile for future use is created in GIS from the well-log data. Any number of cross sectional profiles or 3D images of well-log data are created and viewed interactively. The integration of GIS and BIM is used in the development of 3D visualization of sub-surface pipelines in connection with infrastructure on the earth's surface (Liu and Issa 2012). The results of geo-technical investigations maintained in the GIS database are used to establish the nature of sub-surface rock formations. The profiles derived from well-logs are used for developing a 3D model of sub-surface stratigraphy. This visual representation makes the interpretation of well-logs data more intuitive and reliable manner.

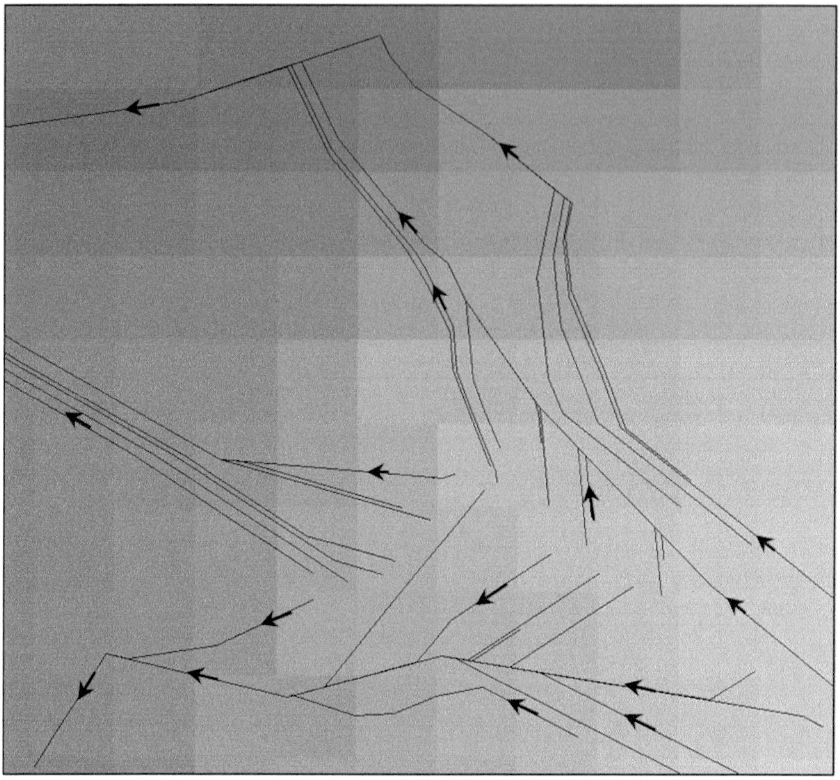

Figure 3.10 Runoff pattern, depicting water flow directions through the steepest path downhill

3.3.4 Modeling Site Constraints and Spatial Relationships

The constraints like all locations must fall exclusively within one particular site boundary, there should not be any overlap between two selected areas, the water supply lines must not cross the sewer lines, etc., are modeled in GIS. The site offices, canteens, material stores, etc., may be constrained to be constructed along existing roads. Different types of infrastructure projects have different categories of constraints in their lifecycle. The modeling of features on the surface of the earth with the help of geometric objects points, lines, and polygons which have spatial relationships with each other is called modeling of topology in GIS. Topology is an approach that integrates the behavior of different features and models different types of spatial relationships between them (Nguyen et al. 2005). Topology describes how various features share their geometry (Borrmann and Rank 2009). It is a mechanism for establishing and maintaining spatial relationships among various features (Borrmann et al. 2009). Modeling topology in GIS is of great use for analyzing spatial relationships with respect to spatial computing perspective. GIS-based topology through a set of validation rules defines how a feature is shared among different activities. Following are a few examples of constrains/relations/analyses modeled in GIS:

- the working area on a site should not belong to two separate activities at the same time,
- calculation of an area served by a crane when installed in a particular location,
- decided locations must be within the particular site area, not on the boundaries,
- endpoints of paths must be covered by a few specific locations,
- specified store locations must be covered by roads, etc.

GIS-based topology provides a user with a set of rules for locations, paths, and areas (LPAs). Site professionals identify the topological constraints and rules that govern spatial relationships among LPAs. The layers are developed in GIS for various LPAs to model the spatial relationships. The layers corresponding to LPAs that are to be used for modeling of spatial relationships are added to the topology. The rules are defined to model the spatial relationships and topology is validated. GIS-based topology analyses, reports, and repairs the spatial integrity after it is initially created as well as after editing. GIS provides a set of topology editing tools which let users to find and fix rule violations. Figure 3.11 depicts the procedure to generate topology, various steps involved are discussed below.

3.3.4.1 *Generate LPAs and Topology*

LPAs to model topological relationships, constraints, and rules that will participate in a topology are generated. For topology validation, LPAs participating and the rules governing their interactions are specified. LPAs participating in the topology are in the same coordinate system. The cluster tolerance is decided that involves fixing a value for snapping feature vertices together to make them coincident. Vertices within a cluster tolerance move slightly during the snapping process. The cluster tolerance value should be small so that only close vertices are snapped together. It should be at least in the order of magnitude smaller than the accuracy of LPAs.

3.3.4.2 *Ranking*

Topology controls which paths or areas are more likely to be moved in the clustering process. For example, when a set of areas is known to have more reliable positions than another set of areas, topology enables less reliable area features to snap

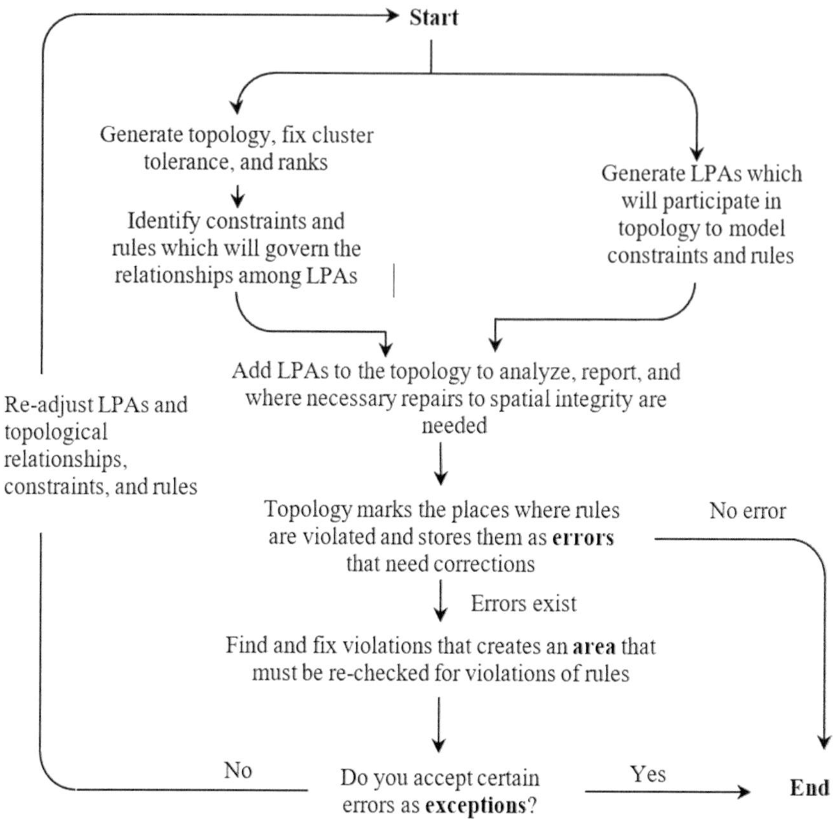

Figure 3.11 Modeling of spatial relationships through topology

to more reliable areas, as shown in Figure 3.12. The vertices of lower-ranked areas within the cluster tolerance are snapped to nearby vertices of higher-ranked areas. Ranks are assigned to areas to accommodate such situations. Vertices of areas that have equal rank lie within the cluster tolerance are geometrically averaged together as shown in Figure 3.13.

3.3.4.3 Assign Topology Rules

Many GIS software tools offer different sets of relationship modeling rules generally called topological rules. Topology rules control the relationships among LPAs within the defined constraints. The initial validation of topology checks all participating LPAs against the defined topological rules. For example, if the constraint, is an area claimed by the two activities at the same time must not overlap,

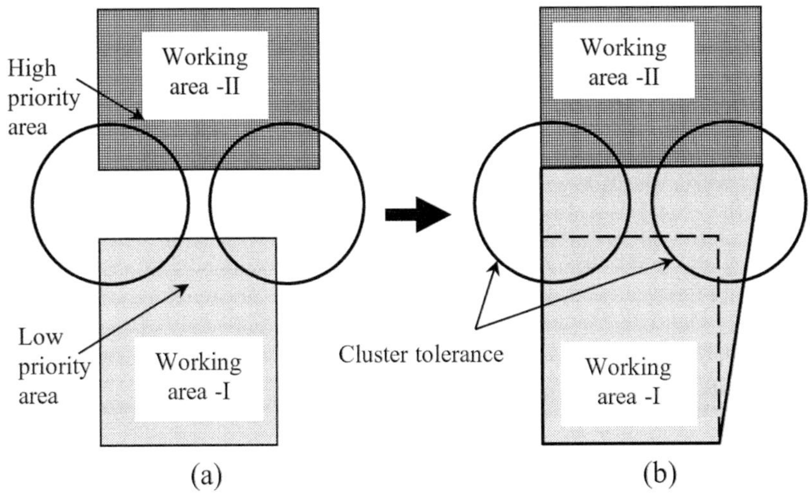

Figure 3.12 Lower priority areas snap to higher priority areas, (a) two unequal priority areas, circles represent cluster tolerances of vertices; (b) interpolated areas where lower priority areas are snapped to higher priority areas

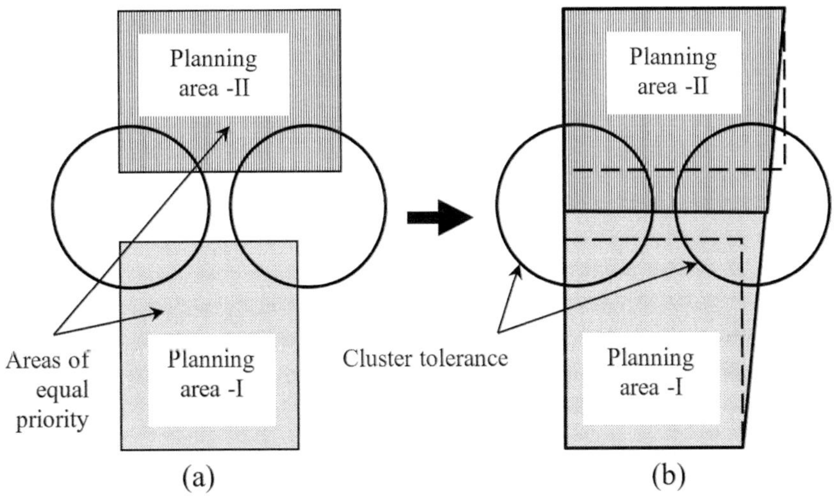

Figure 3.13 Equal priority areas are snapped to the geometric average of their positions, (a) two equal priority areas; (b) interpolated areas

topology checks for the overlaps between areas. Topology marks places where a rule is violated and stores them as errors that need corrections. This section discusses a few topological rules which GIS software tools facilitate (ArcGIS version 9, 2004a, b).

3.3.4.3.1 AREA-BASED RULES

Areas are topologically related to locations, paths, and other areas. Locations may be constrained to fall inside or on the edges of areas. Paths may be constrained to fall along edges of areas, within areas, outside areas, not touching areas, or not crossing the edges of areas. Areas may be constrained not to overlap or allowed to overlap. One area may exactly cover other areas or may not overlap with other areas. The following area-based rules may be used to model the site constraints:

areas must not overlap with each other: used to model the constraints, like area requirement corresponding to one activity must not overlap with the area requirement corresponding to another activity. Areas corresponding to two activities may share edges, vertices, or be completely disjoint. The constraint, that an area cannot belong to two separate activities at one time is modeled with this rule,

areas must not have gaps: used to model the constraint, like there should not be any empty space within a single area or between adjacent areas. This is used to model the constraints like entire areas have been considered in the area planning process or to ensure that the maximum portion of the available area is being planned or utilized,

areas must be covered by other areas: used to model the constraint that a plan must share all of its areas with the areas of another plan. Areas corresponding to the first plan which are not covered by the areas corresponding to the second plan are errors, and

areas must contain locations: used to model the constraint that each area must contain at least one location. Locations must be within an area not on the boundary.

3.3.4.3.2 PATH-BASED RULES

Paths are spatially related to locations, other paths, and areas. Paths are constrained to fall along other paths, connect to other paths at their endpoints, not to touch or intersect themselves, etc. Paths may outline the edges of areas or not touch areas. The following path rules are useful in the spatial modeling of infrastructure projects:

Planned paths must not overlap: used to model the constraint that a path segment should not overlap with another path segment at one time. This is used to model the constraints like the development of the same path should not be in the two different plans,

Path must be covered by a boundary of an area: used to model the constraint like paths should be covered by the boundaries of an area, and

Endpoint of path must be covered by location: used to model the constraint like endpoints of paths must be covered by few specific locations. For example, the path planned must cover specific locations on a site.

3.3.4.3.3 LOCATION-BASED RULES

Locations are topologically related to paths, endpoints of paths, or areas. Locations are constrained to fall along paths, on endpoints of paths, within areas, or on edges of areas, etc. The following location rules are useful:

locations must be inside areas: used to model the constraint like locations must fall within areas,

locations must be covered by endpoints of paths: used to model the constraint like locations must be covered by endpoints of paths, and

locations must be covered by path: used to model constraints like locations must be covered by paths. Used to model the constraint, all major stores should be connected by road/s.

The topology rules mentioned here are limited; this is a never-ending list. However, different types of infrastructure have different types of constraints and need different types of topological rules.

3.3.4.4 Validation of Topology Rules

The topology is validated on GIS-based layers representing LPAs. Topology rules validation does not ensure that rules will never be violated but it informs about violations by marking them as errors. Validation of topology rules checks for any violation of the rules defined. The validation process cracks and clusters the vertices of LPAs. It inserts common coordinate vertices into LPAs that share geometry. The validation generates an error report to make a planner aware of violations and provides an opportunity to repair main errors. Error-prone areas are the places where topology rules are violated. These areas are stored in a topology as a single feature where each new error-prone area is united with the existing error-prone area and each area that has been corrected for errors is removed from the error-prone area. Validation marks editing as error-prone areas, when anywhere editing in LPAs participating in a topology is required to repair main errors. Errors and exceptions allow rendering and management of the places where features do not obey topology rules. Topology stores error features with records where topological errors were discovered during the validation process. Certain errors are acceptable which are marked as exceptions (Bansal 2011).

3.3.4.5 *Topology Editing and Re-Validation*

LPAs participating in topology are edited to repair the topological errors. The topology is re-validated once editing is complete. The areas in which editing/changes have occurred are re-validated or after each editing topology is re-validated. If an alteration in topology rules or ranks is there, it needs to be re-validated again. In this way, relationships between features are established.

3.3.5 Selection of a Sustainable Site

The AEC industry is an essential contributor to infrastructural development. Site selection is the first step in sustainable infrastructure development. GIS helps in the selection of a suitable site for an infrastructure project, keeping sustainability into account. A sustainable site is where an infrastructure facility is developed with the optimum use of resources. A sustainable site must be logistically and economically viable, adequate, safe, and suitable for futuristic expansion. Sustainable site selection is a complex and highly individualistic process. The professionals evaluate various aspects manually based on their judgment, knowledge, and past experience in order to locate an infrastructure sustainably. Site topography, already existing infrastructure facilities, and utilities are the important determinants that ascertain the appropriateness of a site for locating the proposed infrastructure.

In the site selection process, the critical factors affecting the selection of a site for the proposed infrastructure are identified. The rules and regulations that regulate infrastructure site selection by specifying minimum requirements such as open space around existing/proposed infrastructure facilities, distances from existing infrastructure facilities/utilities, maximum slope beyond which no infrastructure facility is to be located, etc., are identified. This helps professionals in making firm decisions for locating infrastructure without damaging the existing natural ecosystem, least damage to nature, harnessing all natural resources optimally, etc. The relation of a location with its surroundings largely influences the decisions of professionals while identifying sustainable sites for developing various infrastructure facilities. Different infrastructure facilities have different site requirements; hence, need different considerations during their site selection.

Identifying a suitable location for an infrastructure is a multi-objective decision-making task that depends upon the critical factors affecting site selection and their relative importance. It is difficult for professionals to simultaneously consider multiple critical factors affecting site selection manually. Various techniques for solving facility location selection problems have been developed. Depending upon the complexity of problems, techniques vary from heuristic to exact ones. Some of the facility location selection problems are solved heuristically on the basis of the experience of professionals. However, for an optimum or appropriate decision, professionals need to support their experience with analytical techniques. Analytical techniques use qualitative and quantitative methods.

Analytical hierarchy process (AHP), fuzzy delphi method, quality function deployment, and analytical network process are some of the qualitative methods used for solving facility location selection problems. The quantitative methods of solving facility location selection problems include empirical selection function, multiple regression, mathematical network flow model, branch-and-bound algorithm, integer programming model, dynamic programming model, non-linear model, goal programming, quadratic programming model, the center of gravity approaches, and GIS. Qualitative and quantitative methods involve multiple criteria in facility location decision-making, hence, known as multi-criteria decision-making methods. GIS-based multi-criteria decision-making methods are used to solve problems that are spatial in nature.

3.3.5.1 GIS in Site Selection

The development of a sustainable lifecycle of an infrastructure has increasingly become a priority in recent years. A sustainable lifecycle starts with suitable site selection. The sustainable site selection of an infrastructure involves the identification of critical factors affecting selection and their degrees of importance or percentage influence. Site selection depends upon the importance of critical factors (locational and topographical) which remain in a continuous state of change. This is due to the change in shape and size of an infrastructure facility, technology, infrastructure requirements, topography, and safety provisions. GIS is used in the site selection to ensure that the critical factors influencing are not overlooked and their percentage of influence is assigned suitably.

The majority of information available in the site selection is not properly used due to the lack of an integrated framework to maintain, manipulate, and analyze the information, thereby failing to locate proposed infrastructure facilities sustainably. Site selection involves large volumes of geo-spatial information which are effectively managed in GIS. Visualization of existing infrastructure facilities/utilities along with their surroundings and geospatial analysis capabilities on a single GIS platform helps in identifying appropriate locations for the proposed infrastructure by using GIS.

Remote sensing images and aerial photographs in GIS provide information for the investigation of infrastructure project site selection. Surface lineament mapping gives prior knowledge of the possible location of faults, joints, weak planes, etc., that may affect site selection. It also helps in geological mapping before ground testing. GIS helps in the selection of sites for rehabilitation and resettlement of project-affected localities by identifying sites having the same environmental conditions to relocate project-affected localities with marginal adverse impact on their traditional means of livelihood.

GIS is used in the site selection of real estate projects, shopping malls (Cheng et al. 2007), large wind turbines, and to assess the impact of infrastructure on landscapes (Kumar and Bansal 2016). GIS with an expert system helps in the analysis of spatial and non-spatial information required in bridge site and bridge type selection (Hammad et al. 1993). GIS-based evaluation being simple and flexible is used to analyze the potential sites for their suitability for the particular type of infrastructure. Site selection of an infrastructure facility in hilly areas where topography plays a major role is done with the help of geo-spatial analysis capabilities available in GIS.

The construction of a dam at a suitable location is a paramount water management approach for flood management and water scarcity relief. The elevation, soil, and geology are essential parameters for the site selection of a dam. The multiple criteria decision-making techniques coupled with GIS are used to solve real-world site selection problems. The analytic hierarchy process and technique for order of preference by similarity to ideal solution (TOPSIS) are two of the upmost extensively used multiple criteria decision-making techniques used for dealing with water-related issues. Analytic hierarchy process and technique for order of preference by similarity to ideal solution approach along with GIS are used for decision making of identifying suitable site of dam considering factors: elevation, slope, drainage density, rainfall, land use and land cover, distance to river, road network density, geology, geomorphology, and soil for a dam site suitability (Pathan et al. 2022).

Water storage by dams helps mitigate drought impacts and ensures water supply for irrigation purposes. GIS is used to identify suitable sites for the construction of dams. The fuzzy analytic hierarchy process and weighted sum method in GIS are used to select suitable sites for dam construction. The factors considered in a dam site selection include lithology, tectonic zones, distance to active faults, distance to lineaments, soil type, land cover, hypsometry, slope gradient, average precipitation, stream width, Curve Number Grid, distance to major roads, distance to towns and cities, and distance to villages (Othman et al. 2020).

Groundwater resources are located based on the digital geological data and the stratigraphic layers of geological maps. Groundwater resources are used in the selection of suitable sites for the construction of subsurface dams. The subsurface dams constitute an affordable and effective method for the sustainable development and management of groundwater resources when constructed on suitable sites. GIS with the support of groundwater balance modeling is used to locate suitable sites for the construction of subsurface dams (Jamali et al. 2013).

The growing use of photovoltaic systems as a result of government incentives and the cost-effectiveness of the technology brings significant environmental benefits and helps in meeting international commitments in terms of renewable energy share. The optimal location selection for a photovoltaic system considers factors such as land use restrictions, topography, environmental, climatic limitations, and proximity to infrastructure. The electric demand and its spatial distribution govern the integration of the photovoltaic system into the main supply for meeting the electric demand. GIS is used for the selection of optimal sites for the construction of photovoltaic plants based on multi-criteria decision-making and spatial overlay with electric load (Zambrano-Asanza et al. 2021).

3.3.5.2 GIS-Based Procedure for Site Selection

A suitable location for an infrastructure facility in an area is the location where maximum requirements are satisfied. In the GIS-based site selection, the available area is narrowed down to a few identifiable locations where maximum requirements are satisfied. The general GIS-based procedure used is shown in Figure 3.14. All critical factors which affect the selection of a site for an infrastructure project are decided from the regulatory requirements and code provisions. The main critical factors are site slope, aspect, elevation, soil type, vegetation, surface runoff, land use and land cover, site accessibility, groundwater table, existing utilities, etc. For example, to identify the relatively flat portion of an area, the slope is derived from the elevation raster. Similarly, to find a site close to/far off the existing facilities/utilities depending upon their importance, Euclidean (straight-line) distances are calculated. GIS is used to develop a layer corresponding to each critical factor to consider it and its relative influence in the site selection process. The number of input layers to be prepared is variable and different for different infrastructure projects.

The value ranges and numbers of classes in different layers corresponding to critical factors are different. For example, the slope derived from the elevation raster may have nine classes. The nine classes of the slope have different value ranges in degrees (units

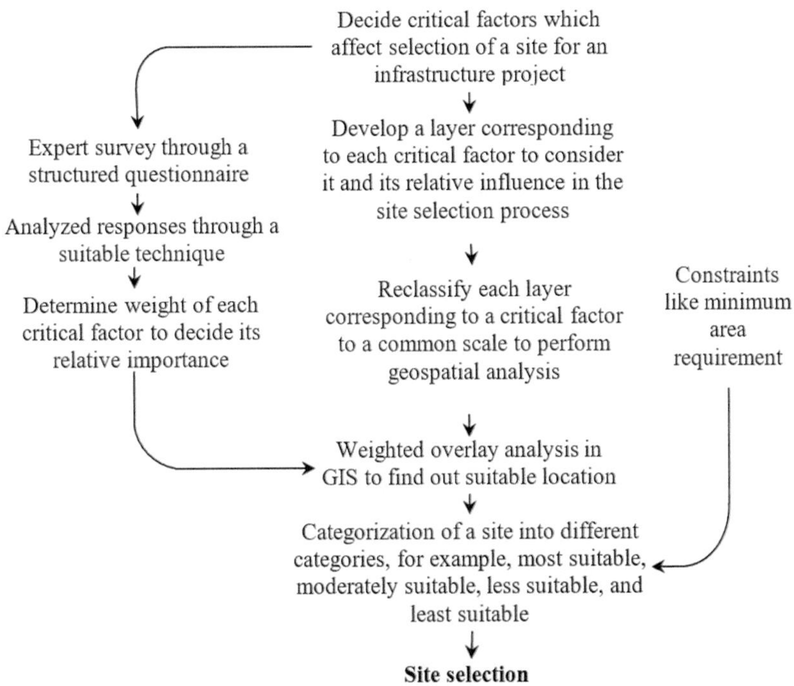

Figure 3.14 GIS-based procedure for site selection of an infrastructure project

of measurement of slope). The numbers of classes are generally kept the same in all layers to perform geospatial analysis. Hence, each layer is reclassified to a common scale, for example, 1 to 4, where 1, 2, 3, and 4 represent most suitable, moderately suitable, less suitable, and least suitable, respectively. For example, the values assigned to the layer corresponding to the slope factor are 1, 2, 3, and 4 for the range of 0° to 20°, 21° to 30°, 31° to 40°, and 41° to 90°, respectively. Various planning regulations are kept in view in deciding ranges in different classes (Kapoor et al. 2019, 2020).

The weight of each critical factor to decide its relative importance in the weighted overlay methods is calculated. Sometimes, an expert survey is done through a structured questionnaire survey and responses are analyzed through a suitable technique to calculate the weight of each critical factor. In other words, the percentage of influence is calculated for each critical factor. The higher the value of weight for a particular factor, the more effective it is in selecting a suitable site. The layer corresponding to each critical factor is used in the weighted overlay analysis in GIS and the derived weight is assigned to it to find out suitable location for the proposed infrastructure facility. This categorizes the site under consideration into different categories: most suitable, moderately suitable, less suitable, and least suitable. Depending upon the availability of areas in different categories, initially site is selected in the most suitable category. However, in case the available area in the most suitable category is less than the required area, the site is selected from the moderately suitable category, and so on (Kapoor and Bansal 2021, 2023).

3.3.6 *Site Access Planning*

Analyzing the accessibility of an infrastructure construction site is important to ensure that construction activities will be performed efficiently. A good construction site accessibility plan reduces possible conflicts during the transportation of heavy equipment and construction materials to a site. GIS helps in planning good accessibility to a construction site. The virtual surface model in GIS is helpful in preparing a good site accessibility plan. Finding the best path from one location to another and virtual navigation along that path have been of great concern in the feasibility stage. GIS is used to develop a virtual navigation system (Souleyrette and Anderson 1998). The development of virtual navigation in GIS has two parts: wayfinding and traveling. Wayfinding involves determining the present location with respect to nearby landmarks/locations and finding a route from the present location to the target location. Traveling is a virtual movement from one location to another in the virtual environment. The geo-virtual navigation system in GIS focuses on wayfinding, virtual traveling, navigation, geo-visualization, and geospatial analyses. The user specifies origin and destination locations in the wayfinding; GIS identifies the path and facilitates virtual traveling along the path computed. Virtual traveling along the path identified is similar to the real-world exploration on the Earth's surface. Virtual 3D geo-spatial representation of an area in GIS repetitively updates positions

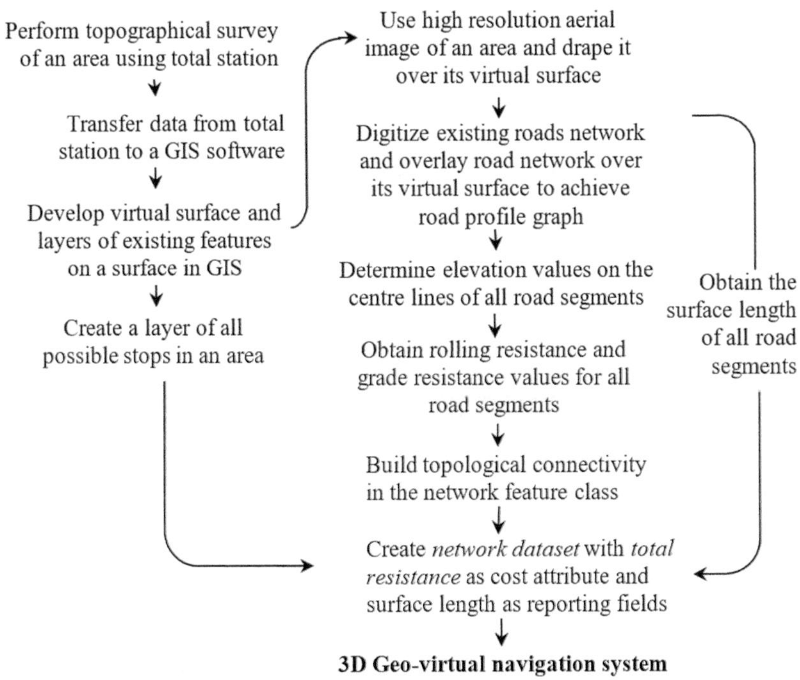

Figure 3.15 Development of a 3D geo-virtual navigation system in GIS

and orientations in space while traveling. The procedure for the development of a 3D geo-virtual navigation system in GIS is shown in Figure 3.15 and discussed in the following sub-sections:

3.3.6.1 Modeling of Attributes

The existing road network is modeled in GIS to find a suitable path between the two locations and facilitate virtual traveling. In GIS, a road network is a system of interconnected edges and connecting junctions. An edge is a linear feature having length through which vehicles move. A junction is a point feature that allows two or more edges to connect and facilitates the transfer of movement of vehicles between various edges. Junctions are used as origins and destinations also. Edges are used to travel from one junction to another. Connectivity among edges and junctions is essential to travel through a road network. An edge must connect to other edges at junctions and the flow from an edge is transferred to other edges through junctions. Both, edges and junctions, have network attributes and properties which control navigation through them.

Network attributes control movement over a road network. Network attributes include time to travel a given length of a road, which streets are restricted for which vehicle classes, speed limits of various roads, streets that are one-way/two-way, etc. The properties are associated with each attribute which are name, data type, units, usage type, etc. How an attribute is to be used during network analysis like cost, descriptor, restriction, or hierarchy is specified during the analysis.

The cost attribute is used to model road impedances. Cost attribute is also called, impedance which are grade resistance, travel time, and distance. In hill areas, grade resistance is considered as a cost attribute for the determination of an optimal route. The cost attribute is applied to all elements of a road network. A road network may have any number of cost attributes; however, network cost is associated with only one cost attribute at a time. Network analysis often involves minimization of the cost attribute during the wayfinding. Descriptors describe characteristics of network elements, for example, the speed limit of streets, the number of lanes, etc. Restrictions are specified for specific elements of a road network. During network analysis, restricted elements cannot be traversed. For example, one-way streets are modeled with the restriction attribute so that they only traverse from one end to the other and not in the reverse direction (Bansal 2023).

3.3.6.2 *Virtual 3D Traveling*

GIS helps in representing geographic features for geo-visualization, geospatial analysis, logistical planning, wayfinding, and virtual 3D traveling. It provides a platform for human-machine interaction where a navigator is able to travel and interact in a virtual 3D geographic space. It facilitates the acquisition of various spatial aspects like distance, area, shortest path, and relative position of various existing features with respect to the current location which is sometimes difficult in real-world exploration. A navigator knows its location in a large geographic area, where other facilities or their locations are, and recognizes what components are being watched. With continuous 3D spatial updating, a navigator while traveling gets information regarding various features in the neighborhood from a current location and feels like moving through the real world. A 3D virtual movement over a landscape along the desired aerial path provides a bird's eye view of an area which facilitates the development of secondary survey knowledge of a navigator.

GIS offers traveling through steering where intermediate viewpoints during the transition are viewed which is quite useful for exploring a site in the feasibility stage. GIS uses steering for virtual traveling on a road and for changing altitude bird's eye view. The availability of on-surface and above-surface virtual traveling makes GIS a suitable tool in the feasibility stage. It is facilitated by changing the locations of a navigator and targets in a space. To view a location from different angles, the location of a navigator is moved. A navigator moves but the center of the view remains fixed. To view different points from the same location, the target's location is moved. The center of a 3D view moves but a navigator remains fixed to a 3D location. This is similar to standing still and swiveling the head to change a view. To move to a new area, the locations of both navigator and target are moved. It is an animation of the movement of a camera along the defined path to look through and experience an area. The camera remains towards the direction of movement, that is, always aligned in the traveling direction. The traveling path rests on the developed surface; therefore, a navigator needs to set a vertical offset above the road to avoid disappearance underneath the surface while traveling. In this way, the level of a camera is fixed at the level of a navigator's eyes. A camera can rotate at any angle to get different perspective views in between. A 3D virtual movement over a landscape along the desired aerial path may be set to reach the chosen target. To fly in the 3D geo-virtual environment both the navigator and target locations are moved. This is similar to looking out the window of a plane as it comes close to landing or flying at a low elevation.

3.3.6.3 *Finding Available Services*

The closest facility analysis in GIS helps in finding the closest facilities such as the closest local market, closest raw materials locations, etc. The network dataset in GIS is used to find the closest facility, display the best route to or from the identified facility, travel cost for the route, and provide driving directions. The cost/impedance cut-off can also be specified beyond which GIS does not search for any facility. The service area analysis in GIS is used to find a service area in a region that encompasses all accessible lengths of roads that are within the specified impedance from the specified location. For example, a five-minute service area for a location includes all roads of a length which can be reached within five minutes from the specified location. It also identifies how much area can be served from a location within the specified time.

3.3.7 *Extending Services to a Site*

The conventional approaches for planning of extending services to a construction site of an infrastructure project involve large amounts of information which are available in various forms. Architects or planners repeatedly retrieve such information from different sources in different forms. A significant portion of effort is wasted on information retrieval and integration. This makes the process time-consuming and prone to errors. GIS is used to integrate information used in route planning to an infrastructure construction site (Varghese and O'Connor 1995). GIS and expert systems are used for automating the tedious and repetitive route-planning tasks for large vehicles within or outside the construction site of an infrastructure. GIS provides data integration, analysis, and display capabilities. The expert system attempts to model the human reasoning process through a set of predetermined rules.

In many infrastructure projects, sites are in close contact/connection with existing public utility lines and structures. Specifically in urban areas, obstacles such as existing public utility lines, railways, canals, and roads influence routes finalizing significantly because of a limited number of feasible crossing points available. The selection of a suitable route to a project site by avoiding existing obstacles not only reduces the risk of damaging existing utilities but also minimizes the project cost and time duration required. GIS is used in the development of systems to automate the process of routing and design of public utility systems to infrastructure project sites by integrating and analyzing the available information (Cheng and Chang 2001). The factors affecting the route selection are identified and represented in the form of layers and attribute tables. The knowledge and data obtained from the experience of experts for route planning are also collected. The surface and underground utilities are represented in several layers to consider their influence on the optimal route selection. In public works, several engineering, construction, and management tasks related to infrastructure are automated using GIS (Adams 1994; Venigalla and Baik 2007).

Restrictions of weights, heights, lengths, and widths of vehicles on the defined route to an infrastructure project location need passing rights or other special permissions. GIS-based automated procedures are used for routing of overweight vehicles carrying loads to an infrastructure project location which needs dimensional and load clearance for bridge crossings on a highway network (Osegueda et al. 1999). GIS helps in identifying all bridges on the specified route and evaluates their adequacy to pass the specified load as well as the required vertical and horizontal clearance. However, many uncertainties and locations of risk on potential routes approaching an infrastructure project location are not modeled. GIS is also utilized in establishing geospatial interdependence between different types of infrastructure networks to avoid duplication of work and disruption of public services (Islam and Moselhi 2012).

Hydroelectric power projects are located in remote and mountainous regions which are difficult to reach. Transportation of large-sized and heavy components like turbines, gates, etc., for such projects constitutes a significant challenge in which GIS has innovative solutions (Varghese and O'Connor 1995). GIS helps in considering the load-carrying capacities of bridges, the turning radius of hairpin bends, roads available, vehicle width and length, etc., in the transportation network analysis. GIS considers all these factors in order to find an optimal route to bring a consignment to a site of an infrastructure project. Based on these and other relevant parameters, the size (micro, mini, small, large, and mega) of a hydroelectric power project is decided. Using a virtual surface model, slope, aspect, catchment area, power generating capacity of a hydroelectric project, etc., are determined. GIS also helps in the rehabilitation or resettlement of affected people.

3.3.8 Route Selection

GIS is used for automated route selection for the construction of water, gas, and power supply lines. The dynamic programming model is used along with GIS for supply route optimization. Environmental restrictions are also taken into account together with operation, maintenance, and equipment installation costs (Monteiro et al. 2005). GIS is used for the selection of optimal routes for water supply pipelines in hill areas (Raj and Bansal 2024).

Electrical energy transmission has become a very important issue with the increase in energy demand. The most important problem in energy transmission is finding the best route in an area. People from different disciplines come together to find the best routes manually. GIS is used in the selection of routes for transmission lines using the least-cost path between two locations (Eroğlu and Aydin 2015). The criteria used in GIS for route selection are slope, landslide, road/railway/pipeline crossing, ice zones, distances to roads, national parks, archaeological areas, residential areas forests, and river crossings, and multicriteria decision methods. The AHP is used to weight criteria relevant to the energy transmission line.

3.3.9 GIS for Environmental Impact Assessment

The use of GIS in EIA has enhanced 3D visualization, process modeling, geospatial analyses, virtual movement, querying, and presentation of outputs in the forms of maps, tables, and graphs. GIS provides access to up-to-date and accurate geospatial data used in EIA (Pham et al. 2007). With an emphasis on using updated geospatial data, the value of the information obtained after geospatial analysis in GIS is far higher than is generally available in text and numeric forms (Satapathy et al. 2008). GIS is an essential component of the EIA process, as the environment is directly affected by the addition or removal of an infrastructure in a locality. GIS software tools are used for developmental planning, policy formulation, decision-making, and in the study of the impact of infrastructure development on natural resources, groundwater, soil erosion, sea-level rise due to global warming, change-detection studies, and delineation of ecologically sensitive areas (Ajman et al. 2021).

Transportation infrastructure plays a vital role in modern society by enabling the movement of people and the transport of goods across local, regional, and international scales. However, while such infrastructure supports economic growth and connectivity, it also introduces significant environmental impacts. These impacts, ranging from noise pollution and air quality degradation to vibrations and habitat disruption, necessitate systematic evaluation and mitigation throughout the infrastructure planning and operational lifecycle. To address these concerns, EIAs are conducted to evaluate the potential environmental consequences of proposed transportation projects. Within this framework, spatial modeling serves as a critical analytical tool for quantifying and visualizing the extent and intensity of environmental impacts across geographic areas. This enables decision-makers, planners, and environmental specialists to assess the implications of different design alternatives and to implement more sustainable solutions.

GIS is widely employed in the environmental evaluation of transportation infrastructure, offering advanced spatial analysis capabilities. One key application of GIS in this domain is the assessment of airborne noise and ground-borne vibration associated with public transportation systems, such as railways, subways, and highways. These disturbances have serious effects on nearby residential areas, public facilities, and ecological habitats. GIS enables the integration of noise and vibration data with geospatial features, such as land use, topography, and population density, to generate impact maps, identify vulnerable zones, and prioritize mitigation measures (Hamed and Effat 2007). By leveraging GIS-based modeling, planners simulate the propagation of noise and vibrations across different terrains, forecast the reach of these disturbances under varying operational conditions, and evaluate the effectiveness of barriers or design modifications. This spatially informed approach enhances the accuracy of environmental predictions and supports more responsible infrastructure planning that balances functionality with ecological and community well-being.

3.3.10 Non-Compliance Checking

The utility regulations stipulate the spatial configurations between underground utilities and their surroundings to avoid interference and disruptions of utility services. Utility compliance checking detects spatial non-compliances in above-ground and underground utilities by examining geospatial data of utilities and their surroundings against textual data of utility regulations. The integration of heterogeneous utility geospatial and textual data for compliance checking is a big challenge (Xu and Cai 2020). The above-ground and underground utility incidents, such as utility conflicts and utility strikes result in time and cost overruns in construction projects, property damages, environmental pollution, personnel injuries, and fatalities (Li and Cai 2015). A main cause of recurrent utility incidents is non-compliance with the spatial configurations between utilities and their surroundings. The utility specifications of the spatial configurations are usually in descriptions in text form. The detection of spatial configurations from the textual descriptions is difficult and time-consuming. This is because of the lack of spatial cognition in many rule-checking systems to process massive amounts of descriptive data. Natural language processing and spatial reasoning automate spatial understanding and utility compliance checking (Li 2016). The natural language processing algorithm translates the textual descriptions of spatial configurations into computer-processable spatial rules. The spatial reasoning executes the extracted spatial rules following a logical order in a GIS to identify non-compliance.

3.4 BIM in the Feasibility Stage

During the feasibility stage, the emphasis lies in assessing the project's viability by analyzing several interrelated variables, which include geographic location, construction costs, associated risks, environmental impacts, etc. Achieving sustainable design of infrastructure projects requires a comprehensive evaluation of these variables and a systematic exploration of all viable alternatives. This stage is critical, as decisions made at this stage significantly influence the project's overall sustainability performance and long-term success. However, selecting the most appropriate design alternative becomes increasingly complex when the decision is dependent on evaluating multiple, conflicting, and design-related variables. The variables such as lifecycle costs, energy consumption, material efficiency, and construction methods must be weighed carefully. This multidimensional analysis is challenging without the aid of sophisticated tools that streamline the evaluation process (Demian and Walters 2014).

To overcome these challenges and improve decision-making, the integration of advanced digital tools and methodologies, most notably BIM and value engineering, has become essential. BIM facilitates the creation of detailed, data-rich 3D models that not only help in visualizing the infrastructure but also embed valuable information related to materials, costs, and environmental performance. Value engineering complements BIM by systematically analyzing project functions to ensure maximum value at the lowest possible cost without compromising on sustainability and quality (OpenAI 2025). Additionally, the analytic hierarchy process serves as a foundational decision-support method in this context. The analytic hierarchy process enables project teams to prioritize and compare design alternatives based on multiple sustainability criteria in a structured and quantitative manner. When used in conjunction with BIM and value engineering, the analytic hierarchy process supports the creation of intelligent models that simulate and evaluate various alternatives. This approach allows stakeholders to identify the most suitable alternatives, based on key sustainability metrics such as cost-efficiency, energy performance, environmental impact of materials, and construction methods (Taher and Elbeltagi 2023). The integration of these tools not only enhances the quality of decisions but also fosters transparency, stakeholder collaboration, and long-term sustainability planning.

The real-time visualization in BIM enables dynamic 3D representations of infrastructure designs as they evolve. Also referred to as real-time rendering or architectural rendering, this capability allows architects, engineers, and stakeholders to interactively visualize how a building or infrastructure will appear and function in a real-world context (Johansson et al. 2015). This not only improves communication among project teams but also supports design validation and iterative decision-making during early project phases. Beyond visualization, BIM significantly enhances indoor navigation capabilities, an area where traditional methods such as 2D floor plans and static routing often fall short, particularly in complex buildings or scenarios involving emergency response, delivery logistics, utility maintenance, and facility management. These limitations are effectively addressed through the semantic and geometric richness of intelligent BIM models, which incorporate detailed spatial data and contextual information. By embedding comprehensive geometric and semantic information, BIM facilitates advanced indoor navigation solutions that are adaptive, accurate, and tailored to specific operational needs (Isikdag et al. 2013).

In summary, the integration of BIM with value engineering, multi-criteria decision-making methods, and real-time visualization not only advances the feasibility assessment process but also empowers project teams to design and plan infrastructure that aligns with environmental, economic, and social sustainability goals. The following sections of this chapter explore the diverse application areas of BIM during the feasibility stage, illustrating how BIM contributes to more informed and environmentally responsible infrastructure development:

3.4.1 Cost-Benefit Analysis

Effectively communicating the potential impacts of an infrastructure construction project to all relevant stakeholders is a critical step in the early stages of project development. In the AEC industry, particularly during the feasibility stage, large-scale public investments often depend on robust decision-support systems to ensure that proposed projects align with strategic national development goals and deliver tangible societal value. One of the most commonly used tools for guiding public project decisions at this stage is cost-benefit analysis. This analytical framework enables the comparison of projected costs against anticipated benefits, thereby supporting more rational and evidence-based decision-making. When cost-benefit analysis is integrated with BIM, it offers even greater value by visually and quantitatively representing project outcomes. The integration of BIM with cost-benefit analysis creates a dynamic and transparent platform that allows stakeholders, such as government agencies, investors, engineers, and the public, to clearly understand the economic, social, and environmental implications of infrastructure projects.

BIM provides a digital representation of the project with detailed data on design, materials, construction processes, and lifecycle performance, while cost-benefit analysis evaluates the financial and non-financial trade-offs. Together, BIM and cost-benefit analysis offer a comprehensive overview of both the short- and long-term impacts of a proposed infrastructure. This integrated approach enhances transparency by presenting complex project information in a standardized and easily interpretable format. It also facilitates better collaboration among multidisciplinary teams, encourages more consistent decision-making processes, and allows for greater automation in the evaluation of design alternatives. As a result, stakeholders are empowered to make well-informed choices backed by accurate data and simulations (Biancardo et al. 2023). Moreover, the synergy between BIM and cost-benefit analysis significantly improves project delivery outcomes across the AEC industry. It supports risk reduction, helps prioritize resource allocation, and contributes to more sustainable infrastructure planning by assessing both quantitative and qualitative impacts. By combining technical modeling with economic evaluation, this integrated approach ensures that infrastructure projects are not only viable but also aligned with broader public policy and development objectives (Omar et al. 2023).

3.4.2 *Deciding Orientations of Buildings*

A well-oriented building significantly reduces energy consumption throughout its lifecycle. Therefore, it is better to decide the orientation of a building in the early stage. Despite the several benefits of proper orientation, current methods and tools for energy simulation face several limitations. The current methods and tools are often time-consuming, difficult for users to understand, and suffer from poor information flows. To address these issues, the integration of BIM and energy simulation tools enables more timely and accurate energy assessments (Kim and Anderson 2011). BIM is increasingly used to assess the impact of building orientation on energy consumption. In a typical workflow, a building is modeled in BIM and then exported to energy simulation software, where various orientation scenarios are tested in the early stage. By analyzing the energy consumption results associated with different orientations, the configuration with the lowest energy demand is identified (Abanda and Byers 2016).

BIM, when integrated with energy simulation tools, plays a critical role in optimizing building orientation, a key factor in enhancing the energy efficiency of the built environment. Building orientation directly influences how much solar radiation a building receives, which in turn affects heating, cooling, and lighting demands. By using BIM in conjunction with energy modeling, designers evaluate various orientation scenarios to determine the most energy-efficient layout for a specific geographic location. These tools allow for simulations that take into account regional climate conditions and site-specific factors, including solar exposure, prevailing winds, and even cultural or social behaviors that influence building usage.

Energy performance is particularly affected by solar heat gain, which varies depending on the angle and intensity of sunlight throughout the day and across seasons. Local factors, such as the social lifestyle of occupants, how and when spaces are used, also contribute significantly to energy usage patterns. Through simulation of different orientation scenarios, designers assess energy consumption impacts early in the feasibility stage, enabling informed decision-making that aligns with sustainability goals.

BIM provides a detailed digital representation of a building, while energy simulation tools use this data to accurately model energy performance. These simulations consider a wide range of parameters, including but not limited to: building orientation, window-to-wall ratios, shading devices, thermal properties of walls and roofs, infiltration rates, lighting efficiency, occupancy behavior, and plug loads performance (Khan et al. 2024). By enabling this analysis, the integration of BIM and energy modeling supports the design of energy-efficient buildings that consume less energy over their lifecycle, reduce operational costs, and minimize environmental impact. This approach is not only instrumental in meeting regulatory standards and certification requirements but also contributes to broader goals of sustainable construction and climate resilience (Yarramsetty et al. 2020).

3.4.3 Sustainability Considerations

The demand within the AEC industry for planning, designing, and constructing sustainable infrastructure is steadily growing to achieve higher performance and cost savings. Sustainability broadly encompasses three key dimensions: environmental, economic, and social well-being. Integrating these sustainability dimensions during the feasibility stage is accomplished through opting for sustainable design, where designers ensure sustainability by selecting appropriate materials and infrastructure systems guided by sustainable rating systems. BIM software tools assist designers in creating 3D building models, with materials and systems chosen from the embedded BIM databases. To address the needs of sustainable buildings, building sustainability assessments are done to evaluate the sustainability of materials and systems. Since BIM contains extensive data about a building's components and materials, it serves as an effective tool for sustainability assessment during the feasibility stage of infrastructure projects (Carvalho et al. 2023). Furthermore, BIM functions as a decision support tool throughout the design process, offering valuable insights to improve building sustainability and help meet decarbonization and climate objectives.

Although BIM has gained significant traction as a widely accepted method to promote sustainability; however, BIM-based tools currently face limitations in quantifying the environmental impacts of materials and systems. This challenge arises from insufficient information on sustainable materials and a lack of interoperability between design and analysis tools necessary for comprehensive lifecycle assessments of buildings. To address this, BIM and lifecycle assessment tools are integrated with specialized databases aimed at sustainable infrastructure project design (Jrade and Jalaei 2013). These integrated databases provide detailed information on sustainable materials, facilitating the creation of sustainable designs and the evaluation of environmental impacts during the feasibility stage. BIM is extensively utilized for lifecycle sustainability assessment and green building planning (Akbari et al. 2024).

BIM evaluates various design alternatives during the feasibility stage. It is utilized to analyze the impact of different design alternatives on a building's energy savings throughout its lifecycle. BIM assists designers in selecting appropriate materials and making crucial decisions that promote sustainability over a building's lifecycle. By integrating with decision-making and problem-solving methods, BIM effectively optimizes the selection of sustainable building components. A decision support system, based on multiple-criteria decision-making techniques, aids in choosing the most suitable sustainable components and design options, considering three key sustainability dimensions. This helps in identifying how design changes affect the overall sustainable performance of the building. The multi-criteria approach within the decision support system uses numerical models to simulate different scenarios, rank alternatives, and select the best options based on strategic priorities and the availability of sustainable materials in the market. The design alternatives generated by the decision support system are then evaluated within an integrated framework that combines the BIM concept with the lifecycle cost method to assess the building's total operational costs.

Designing environmentally friendly infrastructure that delivers both high performance and cost savings is also becoming increasingly important in the development of sustainable cities. Sustainable practices also fulfill our social responsibilities. BIM is integrated with building certification systems, also known as rating systems, to help project teams make informed sustainability decisions while tracking the accumulation of required points according to the chosen green building rating system. With BIM, it is possible to create models that automatically calculate the total certification points earned and the associated registration costs for green and certified materials used in sustainable building design (Jalaei and Jrade 2015). Additionally, BIM enables designers to easily and efficiently visualize and animate sustainable building alternatives in 3D during the feasibility stage (Jalaei et al. 2015).

Sustainability has also become a key focus in the AEC industry due to the significant impact that infrastructure projects have on society, the environment, and the economy. The sustainable lifecycle of an infrastructure aims to minimize negative environmental impacts, reduce construction time, and enhance labor productivity. During the feasibility stage, decisions about construction time and cost are critical factors influencing project outcomes. Traditional manual methods rely on drawings to estimate time and costs, which are labor-intensive and time-consuming, especially when evaluating multiple design alternatives to achieve project sustainability. BIM offers a more efficient solution by streamlining cost estimation and project scheduling. It enables the integration of cost, scheduling, and sustainability considerations early in the project lifecycle (Li 2016).

3.4.4 BIM for Environmental Impact Assessment

The implementation of EIA during the feasibility stage of infrastructure projects in the AEC industry is essential for sustainable development. However, applying EIA in real-world projects is often hindered by a range of challenges. One of the most significant barriers is the fragmented nature of the EIA process, which is typically carried out in isolation across various lifecycle stages of an infrastructure project. This fragmentation leads to inconsistencies in data usage, a lack of coordination among stakeholders, and difficulties in maintaining continuity in environmental evaluation. Additionally, the inherent complexity and technical depth of EIA procedures, involving multiple environmental indicators, regulatory requirements, and simulation tools, often make the process resource-intensive and time-consuming.

To address these challenges, BIM is increasingly being leveraged as an effective tool to support and streamline the EIA process. BIM offers a centralized and data-rich environment that allows for the integration of environmental considerations into the feasibility stage of infrastructure projects. By providing a unified digital platform, BIM facilitates collaboration among architects, environmental engineers, and decision-makers, enabling them to conduct early-stage environmental assessments with greater efficiency and accuracy (Ismaeel and Lotfy 2023). Through real-time simulations and data visualization, BIM-based EIA processes identify potential environmental impacts and mitigation strategies early in the project lifecycle, when design flexibility is highest and changes are most cost-effective.

A key component of environmental performance in the built environment is energy efficiency, particularly in the design of new buildings and major renovations. Improving energy efficiency significantly reduces operational energy demand, thereby lowering greenhouse gas emissions. In this context, the integration of BIM and building energy modeling is essential for the advancement of net-zero energy buildings. These tools enable detailed simulations of energy performance under varying design scenarios, taking into account factors such as insulation, lighting, solar gain, and occupant behavior (Stendahl et al. 2022).

Beyond operational energy, the embodied carbon footprint of construction materials poses a significant environmental concern. Materials like concrete and steel are among the largest contributors to global greenhouse gas emissions due to the carbon-intensive nature of their extraction, manufacturing, and use. As the AEC industry shifts toward more sustainable practices, evaluating the embodied carbon of different material grades has become a critical part in the feasibility stage. BIM enables environmental engineers and project teams to quantify and compare the embodied carbon associated with various structural materials during the feasibility stage. By simulating the environmental impact of alternative material selections, project teams make informed decisions that result in significant reductions in carbon emissions (Shafiq et al. 2015).

Ultimately, the use of BIM to assess embodied carbon demonstrates how early design decisions, particularly material choices, can lead to substantial environmental benefits. This capability highlights the strategic importance of integrating sustainability assessments within BIM workflows from the feasibility stage. As the AEC industry continues to pursue low-carbon construction and environmental accountability, BIM stands out as a critical enabler for achieving meaningful progress toward these goals.

3.5 BIM and GIS Integration in the Feasibility Stage

The integration of BIM and GIS has promising benefits in the feasibility stage but it is a challenge to transfer information from BIM to GIS or vice versa. The software architecture for effective integration of BIM and GIS has been reported in the literature (Kang and Hong 2015). The integration uses the acquisition of data from various sources followed by the transformation of data into an appropriate format; however, the biggest challenge of this method is the large efforts required at an early stage and the isolated development of the ontologies within one particular domain (Liu et al. 2017).

The various aspects of a project, such as environmental assessment, field information, technological validity, and economic feasibility are evaluated in the feasibility stage. Among them, economic feasibility is considered as the most important aspect. The feasibility stage studies consider all costs incurred in the life cycle of an infrastructure project, including land acquisition costs, construction costs, collateral expenses, and operation and maintenance costs. However, only the construction cost is traditionally considered in the assessment of the financial viability of a project (Park et al. 2014). The integration of BIM and GIS is used for estimating different costs of an infrastructure project in the feasibility stage. BIM and GIS integrated model is also used for locating warehouses in large manufacturing factories (Asgari Siahboomy et al. 2021). The applicability of building information models in the geospatial environment in support of site selection and fire response management process has been investigated (Isikdag et al. 2008).

BIM focuses on the micro-level representation of an infrastructure and GIS provides a macro-level representation of the external environments of an infrastructure. Moreover, their integration establishes a comprehensive view of the built environment based on integrated data, which underpins the development and transition of the AEC industry. BIM and GIS integration is used in the sustainable built environment to analyze the status quo and practical applications (Wang et al. 2019).

A fundamental limitation of BIM is its reliance on local coordinate systems, which are typically project-specific and not aligned with real-world geographic coordinates. As a result, the spatial data embedded in BIM do not inherently correspond to their actual positions on the Earth's surface. This misalignment poses significant challenges, particularly for environmental analysis and spatial planning, which often require geospatial context. For example, assessments such as solar radiation, wind exposure, view analysis, shadow casting, and environmental impact studies rely heavily on an infrastructure model's accurate geographic positioning in relation to its surroundings. Because BIM tools are not inherently georeferenced, many of these analyses are typically conducted outside the BIM environment, limiting the efficiency and integration of workflows. To overcome this limitation, automated methods are used to transform BIM data, both geometric and semantic, into their correct geographic coordinates. This process, known as geo-referencing, enables BIM models to be accurately positioned within the geographic coordinate system (Rafiee et al. 2014). Once geo-referenced, the BIM model is integrated with GIS and other spatial datasets, unlocking a wide range of advanced analytical capabilities.

For instance, view analysis and shadow simulation are performed by leveraging the spatial relationships between the BIM model and surrounding geospatial elements, such as terrain models, neighboring buildings, vegetation, and infrastructure networks. These analyses are crucial for optimizing building orientation, minimizing visual obstruction, managing solar access, and ensuring compliance with zoning and urban design regulations. Moreover, geo-referenced BIM plays a pivotal role in urban and regional spatial planning, where multiple infrastructure elements need to be coordinated within a shared geographic framework. By aligning BIM data with real-world geographic locations, planners and decision-makers can better assess the interactions between proposed infrastructure and their surrounding environment, evaluate potential impacts, and design more sustainable and context-sensitive developments. Integrating BIM with geographic coordinate systems significantly extends its utility beyond design and construction. It bridges the gap between building-scale modeling and city-scale planning, enabling comprehensive environmental analysis, improved spatial coordination, and more informed decision-making across the infrastructure lifecycle.

BIM includes very detailed and accurate information about the geometrical aspect of an infrastructure project (Liu and Issa 2012). However, such information is not necessarily georeferenced but uses a local coordinate system. This is hampering many of the environmental analyses in the feasibility stage. Transforming the geometry from BIM to the corresponding location on the surface of the earth in GIS helps in answering many environmental questions efficiently in the feasibility stage. The integration of BIM and GIS transforms the geometrical and semantic information from BIM as a geo-referenced model in GIS (Rafiee et al. 2014).

Building codes are modeled to check a building model automatically based on the codes for non-compliance checking. Many of the building codes are related to the surroundings which makes modeling of surroundings of a building crucial for supporting automated code checking. The urban data to model the surroundings are received from the municipalities. BIM does not contain geographical information about the surroundings of a building and GIS software tools do not model a building and building codes for the compliance checking process. Hence, the integration between BIM and GIS is used to automate code checking. BIM and GIS together are used for the development of a zoning domain model for automated compliance checking of buildings (Altıntaş and Ilal 2021).

The integration of BIM in GIS is currently one of the main investigated research problems, one problem in GIS is the geo-referencing of data imported from BIM. BIM software tools use local coordinate systems or the developed BIM models have no information with respect to their actual geographic locations when imported into GIS. The automatic framework that allows the transformation of the local coordinates of an infrastructure model developed in BIM into its real-world geographic coordinates in GIS was developed by Diakite and Zlatanova (2020).

3.6 Conclusions

The components of the feasibility study in the feasibility stage are project description, technical feasibility, financial feasibility, market analysis, legal and regulatory requirements, risk assessment, and EIA. Although, GIS is not being used in all components of the feasibility study but the chapter discusses many reported application areas in the feasibility stage. The applications of GIS have been reported for database management. GIS deals with both spatial and non-spatial information, hence, the applications of GIS have been reported in the literature for both types of database management. Keeping environmental sustainability in view, nowadays, a proposed infrastructure is modeled along with its surroundings. GIS is being used in the modeling of the surroundings of an infrastructure and the modeling of constraints between an infrastructure and its surroundings. Site selection is the first step in the sustainable development of an infrastructure. GIS helps in the selection of a suitable site for an infrastructure project, keeping sustainability into account. Once a site is finalized, the closest facility analysis in GIS helps in finding the closest facilities, like the closest local market to a site, the closest raw materials locations and their transportation costs, etc., for preliminary cost estimation. GIS helps in the planning for the extension of the facilities and utilities to a site.

BIM also plays a crucial role in the feasibility stage of infrastructure projects by supporting sustainability, cost-effectiveness, and informed decision-making. Through integration with value engineering, multi-criteria decision-making, and cost-benefit analysis, BIM enables systematic evaluation of different possible alternatives based on lifecycle costs, energy efficiency, environmental impacts, and construction methods. Real-time visualization in BIM enhances communication, and integrating BIM with energy simulations optimizes building orientation. Additionally, BIM streamlines EIAs by analyzing embodied carbon, energy efficiency, and materials selection. Overall, BIM enhances collaboration, transparency, and sustainable infrastructure planning from the earliest stage of a project. The integration of BIM and GIS enhances feasibility studies by combining detailed micro-level infrastructure data with macro-level environmental contexts. Despite challenges in data transformation, geo-referencing, and ontology isolation, BIM-GIS supports preliminary cost estimation, site selection, environmental analysis, compliance checking, and sustainable development.

3.7 Exercises

Question 3.1: What is the main objective of conducting the feasibility stage analysis in infrastructure developments?

Question 3.2: How does GIS contribute to evaluating the potential success or failure of a proposed infrastructure project?

Question 3.3: Explain how GIS is used in the development of subsurface profiles from borehole data and analyses of borehole data during the feasibility stage.

Question 3.4: Describe the role of standalone tables in GIS for managing non-spatial information of infrastructure projects, such as labor, equipment, and material costs.

Question 3.5: What is meant by modeling topology in GIS, and why is it important in infrastructure projects? Give examples of area-based topology rules in GIS and explain their relevance in site planning.

Question 3.6: What role does GIS play in sustainable site selection for infrastructure projects?

Question 3.7: Discuss how GIS-based multi-criteria decision-making methods are used in site selections.

Question 3.8: "Validation of topology rules does not ensure rules will never be violated." – Explain this statement with an example.

Question 3.9: Compare qualitative and quantitative techniques in GIS-based site selection. Which approach is more suitable for infrastructure projects and why?

Question 3.10: Compare the elevation raster and triangulated irregular network in terms of accuracy and their applications in surface modeling.

Question 3.11: How does virtual modeling of surroundings in GIS assist planners in analyzing environmental impacts?

Question 3.12: Develop a GIS-based procedure to evaluate the adequacy of bridges along a route for overweight vehicle routing.

Question 3.13: What criteria are considered in GIS-based least-cost path analysis for selecting routes for distribution lines?

Question 3.14: How does GIS enhance EIAs for transportation infrastructure projects?

Question 3.15: What challenges exist in automated compliance checking, and how do natural language processing and spatial reasoning help address them?

Question 3.16: Why is the integration of BIM and GIS important in the feasibility stage?

References

Abanda, F. H., and Byers, L. (2016). "An investigation of the impact of building orientation on energy consumption in a domestic building using emerging BIM (building information modelling)." *Energy*, 97, 517–527.

Adams, T. M. (1994). "GIS-based subsurface data management." *Microcomputers in Civil Engineering*, 9(4), 305–313.

Ajman, N. N., Zainun, N. Y., Sulaiman, N., Khahro, S. H., Ghazali, F. E. M., and Ahmad, M. H. (2021). "Environmental impact assessment (EIA) using geographical information system (GIS): An integrated land suitability analysis of filling stations." *Sustainability*, 13(17), 9859.

Akbari, S., Sheikhkhoshkar, M., Rahimian, F. P., El Haouzi, H. B., Najafi, M., and Talebi, S. (2024). "Sustainability and building information modelling: Integration, research gaps, and future directions." *Automation in Construction*, 163, 105420.

Altıntaş, Y. D., and Ilal, M. E. (2021). "Loose coupling of GIS and BIM data models for automated compliance checking against zoning codes." *Automation in Construction*, 128, 103743.

ArcGIS version 9 (2004a). Introduction part I & II, *Environmental Systems and Research Institute*. N.Y. Street, Redlands, California.

ArcGIS version 9 (2004b). Building a Geodatabase, *Environmental Systems and Research Institute*. N.Y. Street, Redlands, California.

Asgari Siahboomy, M., Sarvari, H., Chan, D. W., Nassereddine, H., and Chen, Z. (2021). "A multi-criteria optimization study for locating industrial warehouses with the integration of BIM and GIS data." *Architectural Engineering and Design Management*, 17(5-6), 478–495.

Bansal, V. K. (2011). "Use of GIS and topology in the identification and resolution of space conflicts." *ASCE Journal of Computing in Civil. Engineering*, 25(2), 159–171.

Bansal, V. K. (2014). "Use of Geographic Information Systems in spatial planning: a case study of an institute campus." *Journal of Computing in Civil Engineering*, 28(4), 1–12.

Bansal, V. K. (2017). "Integrated CAD and GIS-based framework to support construction planning: a case study." *ASCE Journal of Architecture Engineering*, 23(3), 1–13.

Bansal, V. K. (2021). "Integrated framework of BIM and GIS applications to support building lifecycle: a move toward nD modeling." *ASCE Journal of Architecture Engineering*, 27(4), 1–18.

Bansal, V. K. (2023). "A road-based 3D navigation system in GIS, a case study of an institute campus." *International Journal of Applied Geospatial Research (IJAGR)*, 14(1), 1–20.

Bansal, V. K., and Pal, M. (2006). "GIS based projects information system for construction management." *Asian Journal of Civil Engineering (Building and Housing)*, 7(2), 115–124.

Biancardo, S. A., Gesualdi, M., Savastano, D., Intignano, M., Henke, I., and Pagliara, F. (2023). "An innovative framework for integrating cost-benefit analysis (CBA) within building information modeling (BIM)." *Socio-Economic Planning Sciences*, 85, 101495.

Borrmann, A., and Rank, E. (2009). "Topological analysis of 3D building models using a spatial query language." *Advanced Engineering Informatics*, 23, 370–385.

Borrmann, A., Schraufstetter, S., and Rank, E. (2009). "Implementing metric operators of a spatial query language for 3D building models: Octree and B-rep approaches." *Journal of Computing in Civil Engineering*, 23(1), 34–46.

Camp, C. V., and Brown, M. C. (1993). "GIS procedure for developing three-dimensional subsurface profile." *Journal of Computing in Civil Engineering*, 7(3), 296–309.

Carvalho, J. P., Bragança, L., and Mateus, R. (2023). "Automating building sustainability assessment using building information modelling: A case study." *Journal of Building Engineering*, 76, 107228.

Cheng, E. W. L., Li, H., and Yu, L. (2007). "A GIS approach to shopping mall location selection." *Building and Environment*, 42(2), 884–892.

Cheng, M. Y., and Chang, G. L. (2001). "Automating utility route design and planning through GIS." *Automation in Construction*, 10(4), 507–516.

Demian, P., and Walters, D. (2014). "The advantages of information management through building information modelling." *Construction Management and Economics*, 32(12), 1153–1165.

Diakite, A. A., and Zlatanova, S. (2020). "Automatic geo-referencing of BIM in GIS environments using building footprints." *Computers, Environment and Urban Systems*, 80, 101453.

Döllner, J., and Hinrichs, K. (2000). "An object-oriented approach for integrating 3D visualization systems and GIS." *Computers and Geosciences*, 26(1), 67–76.

Eroğlu, H., and Aydin, M. (2015). "Optimization of electrical power transmission lines routing using AHP, fuzzy AHP, and GIS." *Turkish Journal of Electrical Engineering and Computer Sciences*, 23(5), 1418–1430.

Hamed, M., and Effat, W. (2007). "A GIS-based approach for the screening assessment of noise and vibration impacts from transit projects." *Journal of Environmental Management*, 84(3), 305–313.

Hammad, A., Itoh, Y., and Nishido, T. (1993). "Bridge planning using GIS and expert system approach." *Journal of Computing in Civil Engineering*, 7(3), 278–295.

Isikdag, U., underwood, J., and Aouad, G. (2008). "An investigation into the applicability of building information models in geospatial environment in support of site selection and fire response management process." *Advanced Engineering Informatics*, 22(4), 504–519.

Isikdag, U., Zlatanova, S., and Underwood, J. (2013). "A BIM-oriented model for supporting indoor navigation requirements." *Computers, Environment and Urban Systems*, 41, 112–123.

Islam, T., and Moselhi, O. (2012). "Modeling geospatial interdependence for integrated municipal infrastructure." *Journal of Infrastructure Systems*, 18(2), 68–74.

Ismaeel, W. S., and Lotfy, R. A. E. R. (2023). "An integrated building information modelling-based environmental impact assessment framework." *Clean Technologies and Environmental Policy*, 25(4), 1291–1307.

Jalaei, F., and Jrade, A. (2015). "Integrating building information modeling (BIM) and LEED system at the conceptual design stage of sustainable buildings." *Sustainable Cities and Society*, 18, 95–107.

Jalaei, F., Jrade, A., and Nassiri, M. (2015). "Integrating decision support system (DSS) and building information modeling (BIM) to optimize the selection of sustainable building components." *Journal of Information Technology in Construction (ITcon)*, 20(25), 399–420.

Jamali, I. A., Olofsson, B., and Mörtberg, U. (2013). "Locating suitable sites for the construction of subsurface dams using GIS." *Environmental Earth Sciences*, 70, 2511–2525.

Jeljeli, M. N., Russell, J. S., Meyer, H. W. G., and Vonderohe, A. P. (1993). "Potential applications of geographic information systems to construction industry." *Journal of Construction Engineering and Management*, 119(1), 72–86.

Johansson, M., Roupé, M., and Bosch-Sijtsema, P. (2015). "Real-time visualization of building information models (BIM)." *Automation in Construction*, 54, 69–82.

Jrade, A., and Jalaei, F. (2013). "Integrating building information modelling with sustainability to design building projects at the conceptual stage." In *Building Simulation*, 6, 429–444, Heidelberg: Springer Berlin.

Kang, T. W., and Hong, C. H. (2015). "A study on software architecture for effective BIM/GIS-based facility management data integration." *Automation in Construction*, 54, 25–38.

Kapoor, N., and Bansal, V. K. (2021). "Spatial suitability assessment for planning infrastructure facilities at site level in hill areas." *Journal of Urban Management*, 10, 27–45.

Kapoor, N., and Bansal, V. K. (2023). "Planning at site-level in hill regions using geodesign-based approach." *International Review for Spatial Planning and Sustainable Development*, 11(4), 1–18.

Kapoor, N., Bansal, V. K., and Jain, M. (2020). "Development of creative problem solving-based framework for site planning in hill areas." *Frontiers of Architectural Research*, 9(2), 450–466.

Kapoor, N., Jain, M., and Bansal, V. K. (2019). "GIS-based framework for local spatial planning in hill area." *International Review for Spatial Planning and Sustainable Development*, 7(4), 113–130.

Khan, A. M., Tariq, M. A., Alam, Z., Alaloul, W. S., and Waqar, A. (2024). "Optimizing energy efficiency through building orientation and building information modelling (BIM) in diverse terrains: A case study in Pakistan." *Energy*, 311, 133307.

Kim, H., and Anderson, K. (2011). "Energy simulation system using BIM (building information modeling)." In *Computing in civil engineering*, 635–640.

Knoth, L., Mittlböck, M., Vockner, B., Andorfer, M., and Atzl, C. (2019). "Buildings in GI: How to deal with building models in the GIS domain." *Transactions in GIS*, 23(3), 435–449.

Kolbe, T. H., Gröger, G., and Plümer, L. (2005). "CityGML - interoperable access to 3D city models." Proceedings of *International Symposium on Geo-information for Disaster Management*, Delft, Netherlands.

Kumar, S., and Bansal, V. K. (2016). "A GIS-based methodology for safe site selection of a building in a hilly region." *Frontiers of Architectural Research*, 5(1), 39–51.

Kunapo, J., Dasari, G. R., Phoon, K. K., and Tan, T. S. (2005). "Development of a Web-GIS based geotechnical information system." *Journal of Computing in Civil Engineering*, 19(3), 323–327.

Li, H., Yu, L., and Cheng, E. W. L. (2005). "A GIS-based site selection system for real estate projects." *Construction Innovation*, 5(4), 231–241.

Li, J. (2016). *Integrating building information modelling (BIM), cost estimating and scheduling for buildings construction at the conceptual design stage* (Doctoral dissertation, Université d'Ottawa/University of Ottawa).
Li, S., and Cai, H. (2015). "Automated underground utility mapping and compliance checking using NLP-aided spatial reasoning." *Computing in Civil Engineering* 2015, 419–426.
Li, S., Cai, H., and Kamat, V. R. (2016). "Integrating natural language processing and spatial reasoning for utility compliance checking." *Journal of Construction Engineering and Management*, 142(12), 04016074.
Liu, R., and Issa, R. (2012). "3D visualization of sub-surface pipelines in connection with the building utilities: integrating GIS and BIM for facility management." *Proceeding of International Conference on Computing in Civil Engineering*, 341–348.
Liu, X., Wang, X., Wright, G., Cheng, J. C., Li, X., and Liu, R. (2017). "A state-of-the-art review on the integration of Building Information Modeling (BIM) and Geographic Information System (GIS)." *ISPRS International Journal of Geo-Information*, 6(2), 53.
Monteiro, C., Ramírez-Rosado, I. J., Miranda, V., Zorzano-Santamaría, P. J., García-Garrido, E., and Fernández-Jiménez, L. A. (2005). "GIS spatial analysis applied to electric line routing optimization." *IEEE Transactions on Power Delivery*, 20(2), 934–942.
Nguyen, T. H., Oloufa, A. A., and Nassar, K. (2005). "Algorithms for automated deduction of topological information." *Automation in Construction*, 14(1), 59–70.
Oloufa, A. A., Papacostas, C. S., and Espino, R. (1992). "Construction applications of relational databases in three-dimensional GIS." *Journal of Computing in Civil Engineering*, 6(1), 72–84.
Oloufa, A. R., Eltahan, A. A., and Papacostas, C. S. (1994). "Integrated GIS for construction site investigation." *Journal of Construction Engineering and Management*, 120(1), 211–222.
Omar, F., Fateh, M. A. M., and Sitam, S. B. (2023). "Current approach of cost benefit analysis (CBA) into building information modelling (BIM) application: Malaysia government project." In *AIP Conference Proceedings*, 2881(1). AIP Publishing.
OpenAI (2025). ChatGPT (Version 5.1) [Large language model]. https://chat.openai.com/
Osegueda, R., Diaz, A. G., Ashur, S., Melchor, O., Chang, S. H., Carrasco, C., and Kuyumcu, A. (1999). "GIS-based network routing procedures for overweight and oversized vehicles." *Journal of Transportation Engineering*, 125(4), 324–331.
Othman, A. A., Al-Maamar, A. F., Al-Manmi, D. A. M. A., Liesenberg, V., Hasan, S. E., Obaid, A. K., and Al-Quraishi, A. M. F. (2020). "GIS-based modeling for selection of dam sites in the Kurdistan Region, Iraq." *ISPRS International Journal of Geo-Information*, 9(4), 244.
Park, T., Kang, T., Lee, Y., and Seo, K. (2014). "Project cost estimation of national road in preliminary feasibility stage using BIM/GIS platform." *Proceeding of International Conference on Computing in Civil and Building Engineering*, 423–430.
Parsons, R. L., and Frost, J. D. (2000). "Interactive analysis of spatial subsurface data using GIS-based tool." *Journal Computing in Civil Engineering*, 14(4), 215–222.
Pathan, A. I., Agnihotri, P. G., and Patel, D. (2022). "Integrated approach of AHP and TOPSIS (MCDM) techniques with GIS for dam site suitability mapping: a case study of Navsari City, Gujarat, India." *Environmental Earth Sciences*, 81(18), 443.
Pham, T. T., Musy, M., Siret, D., and Teller, J. (2007). "Methodology for integrating and analysing environmental and urban data in 3D GIS." In 10th AGILE *International Conference on Geographic Information Science*, Denmark: Aalborg University.
Rafiee, A., Dias, E., Fruijtier, S., and Scholten, H. (2014). "From BIM to geo-analysis: view coverage and shadow analysis by BIM/GIS integration." *Procedia Environmental Sciences*, 22, 397–402.
Raj, S., and Bansal, V. K.. (2024). "Use of GIS for selection of optimal route for water pipelines in hill areas." *Innovative Infrastructure Solutions*, 9(13), 1–13. https://doi.org/10.1007/s41062-023-01306-3

Satapathy, D. R., Katpatal, Y. B., and Wate, S. R. (2008). "Application of geospatial technologies for environmental impact assessment: an Indian Scenario." *International Journal of Remote Sensing*, 29(2), 355–386.

Seo, J., and Kang, S. (2006). "Geographic information system based roadway construction planning." *Canadian Journal of Civil Engineering*, 33, 508–520.

Shafiq, N., Nurrudin, M. F., Gardezi, S. S. S., and Kamaruzzaman, A. B. (2015). "Carbon footprint assessment of a typical low rise office building in Malaysia using building information modelling (BIM)." *International Journal of Sustainable Building Technology and Urban Development*, 6(3), 157–172.

Sonmez, R., and Uysal, F. (2008). "Geographic information system-based visualization system for planning and monitoring of repetitive construction." *Canadian Journal of Civil Engineering*, 35, 1342–1346.

Souleyrette, R. R., and Anderson, M. D. (1998). "Developing small area network planning models using desktop GIS." *Journal of Urban Planning and Development*, 124(2), 55–71.

Stendahl, M. F., Dubois, M. C., Forgues, D., and Hjelseth, E. (2022). "Building information modeling for environmental impact assessment in early design phases: A literature review." *Open Journal of Applied Sciences*, 12(1), 59–81.

Sun, W., and Hasell, M. J. (2002). "Exploring a GIS prototype to improve the management of the architectural design, engineering and construction building product process." Proceeding of *Environment Science and Research Institute (ESRI) Conference*, http://www.gis.com/library/userconf/proc02/pap0212/p0212 (July 21, 2004).

Taher, A. H., and Elbeltagi, E. E. (2023). "Integrating building information modeling with value engineering to facilitate the selection of building design alternatives considering sustainability." *International Journal of Construction Management*, 23(11), 1886–1901.

Varghese, K., and O'Connor, J. T. (1995). "Routing large vehicles on industrial construction site." *Journal of Construction Engineering and Management*, 121(1), 1–12.

Venigalla, M. M., and Baik, B. H. (2007). "GIS-based engineering management service functions: taking GIS beyond mapping for municipal governments." *Journal of Computing in Civil Engineering*, 21(5), 331–342.

Wang, H., Pan, Y., and Luo, X. (2019). "Integration of BIM and GIS in sustainable built environment: A review and bibliometric analysis." *Automation in Construction*, 103, 41–52.

Xu, X., and Cai, H. (2020). "Semantic approach to compliance checking of underground utilities." *Automation in Construction*, 109, 103006.

Yarramsetty, S., Rohullah, M. S., Sivakumar, M. V. N., and Raj, P. A. (2020). "An investigation on energy consumption in residential building with different orientation: A BIM approach." *Asian Journal of Civil Engineering*, 21, 253–266.

Zambrano-Asanza, S., Quiros-Tortos, J., and Franco, J. F. (2021). "Optimal site selection for photovoltaic power plants using a GIS-based multi-criteria decision making and spatial overlay with electric load." *Renewable and Sustainable Energy Reviews*, 143, 110853.

Chapter 4

The Design Stage

4.1 Learning Objectives

After the completion of this chapter, readers will be able to:

- understand the need and use of GIS in the design stage,
- understand the use of GIS as an alternative for the estimation of construction cost,
- understand the use of GIS for sunlight visualization, and
- understand the application areas of BIM and the integration of BIM and GIS in the design stage.

4.2 Introduction

The design stage refers to the series of sequential steps that are followed in the process of developing the design of an infrastructure. It involves defining the requirements of an infrastructure project, its physical constraints, and the allocated budget. It also involves conducting the analysis for developing the conceptual design. The design team is decided to develop broad concepts and ideas for the conceptual design. Design is presented in the form of sketches, diagrams, and concepts to explore different possibilities, out of which one is finalized. The design team translates the finalized concepts into more detailed designs and drawings. Finally, the developed design is further refined and detailed in the form of architectural and structural designs. The specifications and materials that have a limited or no impact on the environment are decided and the design is developed to a level where it can be used for the construction of an infrastructure. At this stage, if it is a building infrastructure, components of a building, such as shading devices, window orientation, daylighting, and landscape, are also designed. Primarily, this stage involves the preparation of architectural design, structural design, preparation of detailed working drawings, and deciding specifications that contractors use for deciding the bid price and executing construction. The design drawings mainly include details about site layout, architecture, structure, working, mechanical, electrical, and plumbing systems. The contractors submit their bids based on the design drawings and specifications.

DOI: 10.1201/9781003759348-4

With geographic information system (GIS), it is possible to simulate different flood conditions in the surroundings of an infrastructure. This helps in developing a design that has a minimum possibility of flooding and other types of failures to the infrastructure. A 3D model of an infrastructure along with its surroundings is developed to assess the aesthetic appeal of the design and sustainable decision-making. Accessibility is another important aspect of the design, once a project site is finalized, access is designed using GIS. The gradient of the access road is governed by the nature of heavy equipment that are to be transported to the project site in which applications of GIS are widely reported. The design of the road alignments is facilitated by using digital elevation model (DEM) in GIS which helps in the calculation of the values of slopes and elevations. Site analysis in GIS helps in deciding a better orientation for an infrastructure, easy storm water drainage from the site, design with respect to wind direction, minimizing disaster risks involved, and planning for scenic views. This chapter, in the following sections, covers the application areas of GIS, building information modeling (BIM), and integration of BIM and GIS in the design stage of infrastructure projects.

4.3 GIS in the Design Stage

The design stage develops infrastructure with respect to architectural and structural contexts. GIS aids in road alignment designs, cost estimation, sunlight visualization, and energy analysis, which are discussed in the following sections:

4.3.1 *Public Participation in Design*

Public participation is necessary in the design stage in order to develop the sustainable lifecycle of an infrastructure project. GIS is used to develop an online platform that involves the public in the design stage for decision-making. A GIS-based visualization technique is used to enhance public participation in the design stage (Al-Kodmany 1999). GIS is an effective tool to visualize and simulate the complex topographical conditions of a construction site, like a dam construction during the design stage (Zhong et al. 2004).

4.3.2 *Alignment Design*

Sustainable road alignment planning in built-up areas struggles to meet the traffic demands of society within the limited spaces available for road construction. Sustainable road alignment planning in built-up areas is affected by technical constraints, traffic, economic conditions, social, and environmental factors. Road alignment is significantly influenced by existing buildings, road networks, land use, and noise and air pollution impact on residents (Jiang et al. 2022). In addition, road width and road widening are significant factors in road alignment planning. In GIS, the least-cost path algorithm is employed for sustainable road alignment planning in the built environment in which building demolition, land use, traffic congestion, noise impact, air pollution impact, and construction costs are considered. GIS is used to consider sustainable factors with different weights for generating various road alignments from different perspectives.

4.3.3 GIS in Construction Cost Estimation

The computer-based analyses have changed the way construction cost estimates of infrastructure projects are produced and continue to change as new software tools are being developed. Electronic digitizers to trace drawings are used in the automated quantity takeoff process. Electronic spreadsheets are also used for the quantity takeoffs and pricing sheets. Some of the commercially available cost estimating software tools take quantities determined either manually, from CAD files, or by using digitizers and apply a database of unit prices to calculate the construction cost of an infrastructure project. BIM or CAD-based systems are also capable of generating quantity takeoffs and materials requirements, which are used to find the cost of construction. The cost estimates prepared without detailed engineering data of an organization are less accurate. GIS is used for quantity takeoffs and cost estimation from the data of an organization itself. GIS helps in determining the costs of various resources required. The GIS-based approach for quantity takeoffs is accurate enough and eliminates the possibility of errors like missing or duplicating different items of work involved.

4.3.3.1 *Quantity Takeoffs*

The design drawing layers developed in CAD-based systems or GIS are used as the spatial data representation of the architectural drawings in GIS (Cheng and Yang 2001). The CAD-based systems are generally used in the AEC industry for developing architectural drawings because of their user-friendly editing capabilities. GIS extracts necessary dimensions from the layers to perform various calculations for quantity takeoffs. Accurate bills of quantities (BOQ) are generated in GIS on the basis of the dimensions of various components represented in different layers. Each layer in GIS corresponds to one or a part of an item of work whose quantity is to be estimated. The number of layers to be developed for each item of work depends on the shape, openings, and thickness at different levels of height. The main attributes used for quantity takeoffs in GIS are the length, perimeter, and area of various geometric objects in a layer. In general, all layers are developed using line or polygon objects. The types of objects used to construct different layers depend upon the unit of measurement of various items of work. For example, for the units *running meter* and *square meter* of the items, a geometric object line may be used as shown in Figure 4.1. For the units *square meter* and *cubic meter* of the items, a geometric object polygon may be used to create the data layer. The GIS-based quantity takeoffs procedure eliminates errors like missing or duplication of various items of work by visualizing each component corresponding to the items of work (Bansal and Pal 2007).

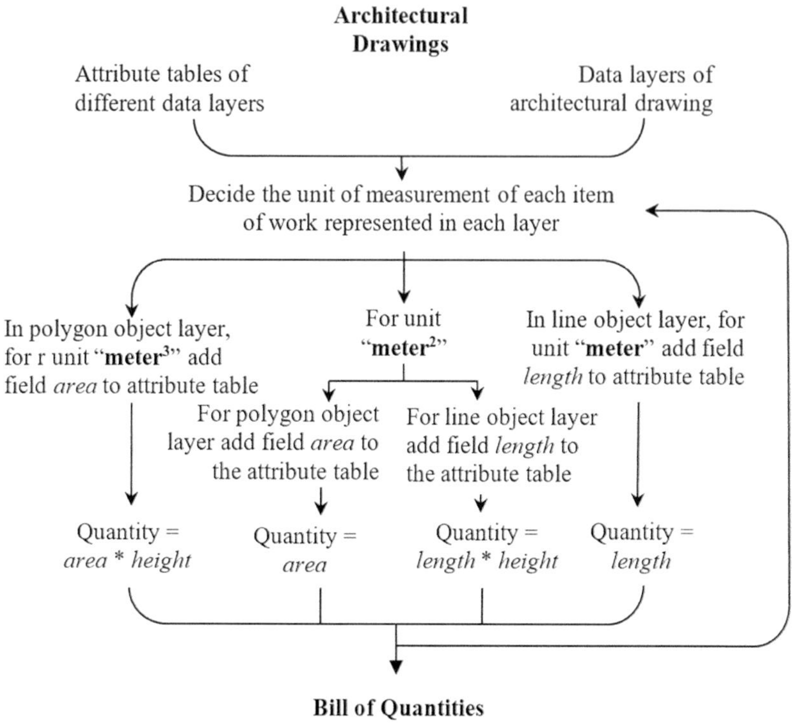

Figure 4.1 Flow diagram to generate a bill of quantities

4.3.3.2 *Cost Estimation*

To increase the accuracy of cost estimates, GIS uses well-defined engineering data related to the organization itself. The cost estimate produced from organizational data usually gives a better estimate than any published data because it includes a group of tasks such as downtime, cleanup times, lunch, and tea breaks which vary from organization to organization. The database management capabilities of GIS are used to store resource data which are used in the cost estimation. The information about construction materials, different types of workers to be employed, and equipment needed corresponding to various items of work are stored in different standalone resource tables in GIS. The costs of different items of work, represented in the form of different layers in GIS, are obtained by linking the BOQ with resource tables in the database. Different elements used for cost estimation are materials requirements, workers needed, and equipment to be used. For any infrastructure construction project, the total cost may contain one, two, or all of these three elements. All three elements of cost are computed by linking the BOQ with the three resource tables in the database. Each element is priced separately in accordance with the unit of measurement corresponding to items of work in each row of the BOQ. The costs of the three elements are added to compute the total cost of construction.

For example, to generate the bill of materials (BOM), a standalone table in GIS is used to store the information regarding various types of materials required for a unit of all possible items of work. BOQ is used as the input; GIS extracts the materials required for different items of work in the BOQ. BOM is generated; it has the field corresponding to each material required and the cost associated. Finally, the total cost of materials for different items of work, represented by various data layers, is calculated. Similarly, requirements of workers and equipment corresponding to each item of work are generated and their costs are calculated.

GIS-based tools are used for cost estimation of road infrastructure, in which the conditions of the road and the cost of maintenance of roads are taken as inputs. The database is linked with interactive GIS-based maps to facilitate visualization of road infrastructure and their construction cost estimation (Shrestha and Shrestha 2014).

4.3.3.3 Site Specific Costs

Generally, conceptual cost estimates are usually prepared through the aid of historical data and later adjusted to take into account project-specific characteristics such as location. The common practice for adjusting cost estimates for a location is to use a set of location cost factors. When a location is not represented in the chosen data set, the nearest location factor is usually interpolated. The validity of this interpolation method is not substantiated. The GIS-based technique is used to conduct spatial analysis of the cost index reference data for spatial autocorrelation (Migliaccio et al. 2009, 2015).

The costs of materials, labor, and equipment are used in the cost estimation to calculate the total cost of construction. However, the cost of construction varies from one location to another location due to the availability of materials, workers, equipment, and transportation costs involved. Hence, GIS-based tools for the computation of materials costs on site are used, which consider the distances from supply points and the topography along the routes to find the costs on a particular site.

4.3.4 Sunlight Visualization

The total energy consumption is minimized to plan the sustainable lifecycle of an infrastructure. Deciding the location and orientation of an infrastructure provides opportunities to reduce energy consumption during the design stage. A well-located and oriented infrastructure relates better to its site as well as the climate. It maximizes the opportunities for passive solar heat gain during winters and solar heat gain avoidance during summers. The visualization tools help designers understand and recognize solar heat gain/avoidance during the design stage.

The sun's lighting occurs in 3D space; in addition to this, the sun's position fluctuates with time (diurnal and annual cycle of a year). Therefore, to preview sun lighting on an architectural space over time, a designer navigates in three dimensions of space and two dimensions of time. GIS is used to develop navigable animations of sun lighting in three dimensions of space and two dimensions of time. GIS uses local time and location (latitude and longitude) as input to calculate azimuth and altitude angles of the sun (Bansal and Pal 2009). The sun angles are used in GIS to render the virtual surfaces with colors of varied gradients depending upon the intensity of sunlight received. The designers may change their view positions/locations to examine the sunlight intensity received on different faces and sides of buildings. The animations are developed to illustrate the time-dependent phenomenon of sunlight in a navigable 3D space in GIS.

An animation is an effective tool to demonstrate the variations of sunlight intensity on the surface of an infrastructure over annual and diurnal periods. GIS provides the facility of exporting animations to other platforms which are independently used for demonstration. Figure 4.2 shows navigable 3D views of a sample building, with a side parallel to the East–West direction, at different times. Figure 4.3 shows how a horizontal surface receives more solar heat gain in the

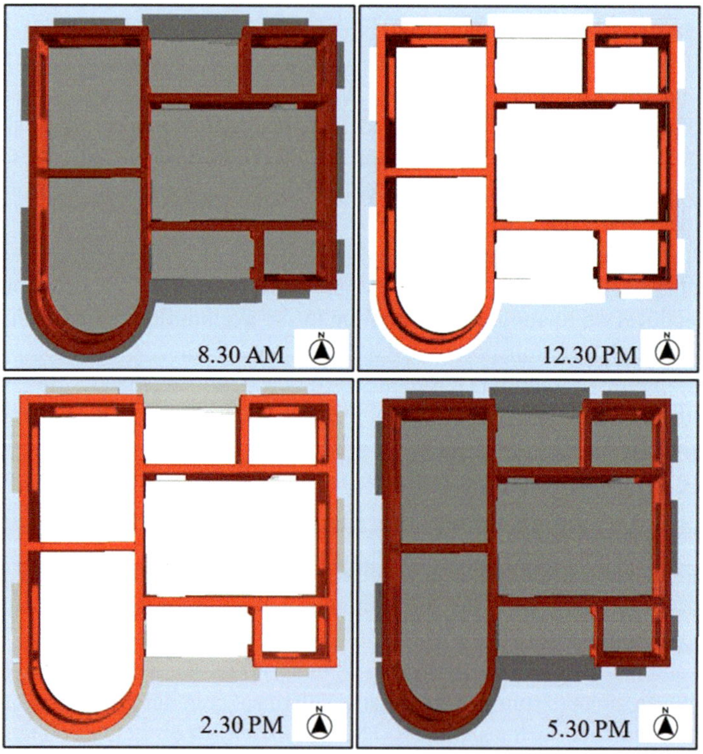

Figure 4.2 Navigable views of a sample building at different times of the day, the building has a side parallel to the east–west direction

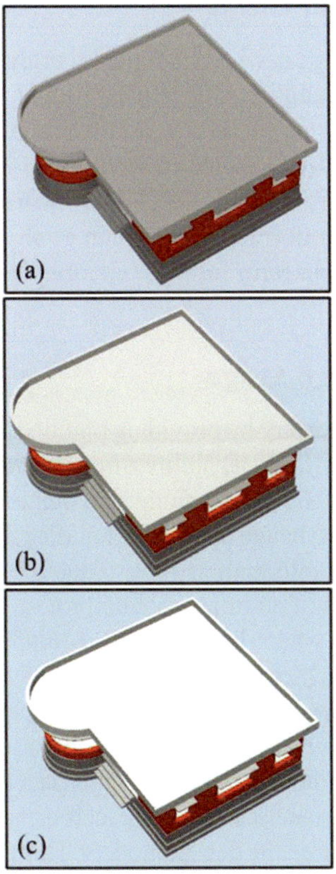

Figure 4.3 Navigable 3D views of a sample building that depict horizontal surfaces receive more solar heat in summers than in winters in the northern hemisphere: (a) on 6 December, (b) on 3 April, and (c) on 15 May

summer than in the winter at the same location in the northern hemisphere. The optimal orientation of an infrastructure also depends upon the purpose of its use. For example, to use solar energy during winter for passive solar heating, south-facing glazing is desired. The sun shading is relatively easy on south-facing windows, which helps in minimizing the solar heat gain during summers also whereas the north-facing wall receives very little solar heat gain. The sun shading on east and west-facing glazing is quite difficult because of low sun angles and contributes to unwanted heat gain on buildings. An infrastructure project is located with respect to nearby existing facilities and utilities, where the sunlight visualization becomes a geospatial problem (Mardaljevic and Rylatt 2003; Rylatt et al. 2001). GIS is also utilized to explore the effects of urban texture on buildings' energy consumption based on the analysis of DEM of cities (Ratti et al. 2005). GIS is used for evaluating the total annual/monthly irradiation incident on building facades in urban settings (Rylatt et al. 2001).

4.3.5 *Estimation of Solar Energy*

The solar energy in buildings is widely adopted worldwide to reduce carbon emissions for enhancing sustainability. GIS is used to find the amount of solar energy that may be produced in a city or an area and what proportion of building energy requirements may be fulfilled by solar energy. A GIS-based energy balance modeling system for urban solar buildings that extends from a single building to the urban building system was developed by Quan et al. (2015). It uses urban-scale data through a GIS-based platform for solar energy planning.

4.4 BIM in the Design Stage

BIM transforms the AEC industry by providing highly detailed 3D visualizations that go beyond simple drawings. This capability offers a shared and standardized digital environment where all stakeholders (architects, engineers, contractors, and facility managers) can access and exchange up-to-date building information seamlessly. This unified platform eliminates information silos, reduces errors, and fosters collaboration from the earliest stages of an infrastructure project. With the rise of smart built environment technologies, where buildings are equipped with embedded sensors, automated controls, and Internet of Things devices, BIM's role becomes even more vital. These smart objects are designed to collect and transmit real-time data to optimize building operations such as lighting, HVAC, security, and energy management. BIM acts as the central communication hub for this data flow, enabling designers and facility managers to decide materials, detect defects early, and integrate smart device control parameters directly into the design models. This integration results in smarter buildings that are more responsive to occupant needs and operational demands.

BIM facilitates automated structural analysis by generating complex models of structural geometry, load resistance, and material performance based on interconnected data streams, including architectural designs, construction schedules, and material specifications. This interlinked approach not only improves precision but also accelerates the iterative design process, enabling engineers to assess the structural integrity and optimize resource use quickly. Beyond design optimization, BIM supports automated quality assurance through rule-based checking systems that verify compliance with building codes, sustainability standards, and client requirements (Choi et al. 2020). This systematized approach enhances construction quality by catching potential issues before they manifest on-site, thus reducing costly rework and delays.

One of BIM's greatest strengths lies in its ability to enhance collaboration across the design team. Effective BIM implementation relies on established protocols, structured sets of procedures, flowcharts, and communication standards that ensure smooth data exchange and decision-making among architects, engineers, and clients during the design stage. These protocols help coordinate complex workflows, maintain data integrity, and enforce compliance with technical and regulatory requirements during the design process (Kassem et al. 2014). The abstraction of these protocols into BIM models allows project managers to monitor the relationships between processes and technologies, ensuring that every stage, from conceptual design to construction, adheres to best practices and regulatory frameworks (OpenAI 2025). This structured collaboration is essential to managing increasingly complex projects, reducing risks, and achieving timely delivery.

The integration of BIM with artificial intelligence marks a transformative evolution in the AEC industry. The industry faces ongoing challenges, including low productivity, budget overruns, workforce skill gaps, and the complexity of managing vast amounts of data across project stages. Digital fabrication, smart devices, and automation offer solutions, but when combined with artificial intelligence-enhanced BIM, they fundamentally reshape the industry workflows. Artificial intelligence technologies, especially machine learning, utilize statistical algorithms to enable systems to learn from historical data, identify patterns, and make data-driven predictions. When embedded within BIM platforms, artificial intelligence predicts building system performance under different conditions, detects potential design conflicts or inefficiencies early, optimizes building operations by simulating alternative scenarios, and enhances decision support by prioritizing design alternatives based on multiple criteria such as cost, sustainability, and user comfort. The successful integration of artificial intelligence with BIM requires addressing technical challenges like data interoperability, software compatibility, and alignment with project goals. The potential benefits are enormous, including enhanced design accuracy, reduced waste, improved sustainability outcomes, and smarter infrastructure management throughout the building's lifecycle stages.

As sustainability increasingly becomes a non-negotiable requirement, BIM enables extensive building performance analyses during the design phase. These analyses encompass energy and thermal modeling to simulate building energy use and thermal comfort, green building rating analysis to assess certification compliance, cost-benefit analysis to evaluate economic and environmental viability, wind and ventilation studies to ensure airflow and reduce reliance on mechanical systems, acoustic analysis for noise control, and water efficiency analysis to optimize plumbing systems (Chang and Hsieh 2020). By integrating these varied performance metrics into a single model, BIM empowers stakeholders to make holistic, informed decisions that balance functionality, cost, and sustainability. Public involvement is becoming an essential part of sustainable urban development. BIM integrated with community engagement platforms, ranging from traditional consultations to modern crowdsourced design initiatives, facilitates transparency and broader participation (Grover et al. 2015). By visualizing projects in accessible 3D formats and linking BIM data with social media, stakeholders gather valuable feedback and build consensus, ultimately leading to designs that better reflect community values and expectations.

Sustainability certification systems such as *Leadership in Energy and Environmental Design* (LEED), *Building Research Establishment Environmental Assessment Method* (BREEAM), *WELL Building Standard*, and others require comprehensive assessments covering diverse building aspects, architectural form, construction materials, site context, and mechanical, electrical, and plumbing systems. Traditionally, these evaluations involve labor-intensive and fragmented processes. BIM revolutionizes this by enabling multi-disciplinary data integration within a single digital model, allowing for rapid and accurate documentation needed for certification (Azhar et al. 2011). This integrated approach not only streamlines certification but also encourages iterative design improvements by providing real-time feedback on sustainability performance. Designers explore trade-offs between cost, material choices, energy efficiency, and environmental impact, enabling them to develop buildings that meet or exceed certification criteria more efficiently.

BIM serves as a foundational technology that enhances every facet of modern building design. From enabling real-time 3D visualization and integrated performance analyses to fostering collaborative workflows and supporting smart building systems, BIM's role continues to expand in response to evolving industry demands. The integration of artificial intelligence and machine learning further amplifies BIM's capabilities, presenting unprecedented opportunities to optimize infrastructure for sustainability, efficiency, and resilience in the face of complex environmental and economic challenges. The following sections explore the diverse application areas of BIM during the design stage, illustrating how BIM contributes to more informed and environmentally responsible infrastructure design:

4.4.1 Building Design

The implementation of BIM in the design stage is fundamentally transforming traditional structural design workflows. Both academia and the AEC industry increasingly recognize BIM as a powerful tool for reshaping the way structural systems are conceptualized, analyzed, and optimized. BIM facilitates comprehensive modeling, structural analysis, and performance optimization by offering a systematic approach to design through intelligent data modeling, interactive visualizations, and standardized data exchange formats. Its use has a notable influence on the roles and responsibilities of personnel involved in the structural design (Chi et al. 2015). BIM significantly contributes to reducing design errors and minimizing rework in infrastructure construction projects. BIM enables designers to better understand the root causes of design errors both in terms of why they occur and how to prevent them (Love et al. 2011). Furthermore, BIM provides design teams with actionable and function-specific guidance aimed at defect prevention during the design stage (Koo and O'Connor 2022). The BIM-based design process is instrumental in identifying the essential data, stakeholders, tasks, and BIM functionalities required to proactively avoid common design defects.

Selecting optimal building components is central to achieving the multifaceted goals of sustainable construction. BIM plays a vital role in this process by facilitating informed decision-making during the design phase. It is used to evaluate entire building systems or specific components to visualize optimal solutions using 3D or 4D models. This capability supports a more data-driven and integrative design strategy (Inyim et al. 2015). When assessing the structural integrity of buildings subjected to seismic forces, accurate identification of damage modes and severity is critical. These assessments are often labor-intensive and prone to human error. BIM aids in this domain by encapsulating damage-related data alongside geometric and structural information. With the support of reasoning algorithms and transformation tools, BIM enables automated strength analysis and visual assessments of reinforced concrete structures (Anil et al. 2016).

The integration of BIM with finite element analysis software has further advanced the structural analysis of buildings, bridges, and industrial structures. This integration enables more detailed and accurate modeling by converting BIM representations into suitable geometric and material definitions for finite element analysis. In this approach, structural elements represented as solids in BIM are abstracted into beam and shell elements based on their geometry. Material behavior, such as concrete damage plasticity or steel plasticity, is then simulated to assess structural performance under load. High-performance computing environments are typically utilized to handle these complex simulations, especially in the case of high-rise buildings (Ren et al. 2019).

Additionally, the combination of BIM and augmented reality technologies offers new avenues for optimizing design workflows and minimizing errors. This integration ensures real-time data accessibility and enhances communication and collaboration across project teams. Despite challenges in technology integration, data interoperability, and user adaptation, BIM-augmented reality solutions significantly enhance project visualization and streamline coordination, driving the industry toward more sustainable and efficient practices (Bhatarai et al. 2025).

Sustainable design has become a mainstream objective in building infrastructure projects, with sustainable development being prioritized more than ever due to the intensifying impacts of climate change. BIM enhances sustainable building design by enabling better information management and performance analysis throughout the design process. Its core contributions to sustainable design include promoting integrated project delivery and supporting iterative design optimization (Wong and Fan 2013). By facilitating early-stage decision-making, BIM helps reduce costs associated with design changes and supports the selection of energy-efficient and resource-conserving solutions. As sustainable building practices gain prominence in response to global climate change and the need to reduce greenhouse gas emissions, BIM's role in promoting environmentally responsible design becomes even more critical. BIM-based optimization methods support the development of sustainable building solutions by helping designers identify optimal design alternatives. Techniques such as particle swarm optimization are applied to explore trade-offs between lifecycle costs and carbon emissions, enabling broader design exploration and faster decision-making processes (Liu et al. 2015). This supports the development of economically viable and environmentally friendly building designs.

Environmental impact assessments are gaining importance in both public and private sector projects due to the long-term ecological implications of design decisions. However, conventional environmental impact assessment processes are typically done at the end of the design, when changes are more difficult and costly to implement. BIM offers a solution by integrating environmental impact assessments into the early stages of the design process. Despite existing challenges, such as a lack of standardized data structures and limited interoperability between BIM and environmental impact assessment software, efforts are being made to establish bi-directional data exchange and seamless integration. For instance, a BIM-based environmental impact assessment framework that employs a structured data schema allows real-time environmental impact assessments to be executed during the design stage, providing immediate feedback on the environmental impact of different design choices (van Eldik et al. 2020).

4.4.2 Design of Energy Efficient Buildings

The building sector is a major contributor to global environmental challenges, accounting for approximately 32% of the world's total energy consumption and 19% of all energy-related greenhouse gas emissions (Xu et al. 2021). These figures underscore the urgent need for energy conservation measures and greenhouse gas emission reductions within the building industry. As urbanization intensifies and the demand for buildings increases, it becomes increasingly important to adopt innovative solutions that address energy inefficiency and support environmental sustainability. One of the most transformative tools in this context is BIM, which is especially effective in addressing the gap between projected (designed) and actual energy performance in buildings. This discrepancy often arises due to the lack of integration between design intentions and real-world performance parameters. BIM helps mitigate this gap by facilitating early-stage energy analysis, design reviews, and energy-related quality control processes.

In light of the environmental degradation caused by high energy demand in buildings, the efficient use of available energy resources has become a critical priority. BIM supports this goal by enhancing a building's overall energy efficiency and sustainability performance. For instance, BIM allows for detailed analysis of natural lighting conditions, thermal comfort zones, and energy consumption patterns, enabling the design of buildings that leverage daylight and natural ventilation to reduce reliance on artificial lighting and mechanical cooling or heating systems. This not only results in energy savings but also contributes to improved indoor environmental quality, which plays a role in enhancing occupants' health and reducing the transmission of airborne pathogens (de Lima Montenegro et al. 2021).

The situation becomes even more critical in the case of industrial buildings, which typically exhibit higher energy demands due to large-scale mechanical systems, production equipment, and longer operating hours. Industrial buildings require advanced modeling and simulation tools that can identify synergies between architectural design, building services, and manufacturing systems. BIM serves as a centralized knowledge repository, allowing data from architectural, structural, mechanical, and production models to be exchanged and analyzed seamlessly. This integrated approach enables follow-up analyses, such as cost estimation, structural integrity assessments, and thermal behavior simulations, supporting more holistic and optimized building solutions (Gourlis and Kovacic 2017).

With the growing global awareness of climate change and the implementation of increasingly rigorous building codes and sustainability standards, architects and engineers are now placing greater emphasis on energy performance from the earliest stages of design. However, in conventional architectural practice, building performance simulations are typically conducted after the design is completed, limiting their potential to influence fundamental design decisions. To address this, BIM enables the seamless integration of building materials, components, and technical systems into the initial design stage, thereby allowing comprehensive performance evaluations to occur concurrently with design development. The multi-disciplinary nature of BIM is a key advantage, as it allows access to diverse data sets required for accurate energy modeling. Parameters such as thermal properties, daylighting potential, heat ventilation and air conditioning system efficiency, and envelope characteristics are analyzed in real-time. BIM supports instantaneous energy calculations and visually intuitive performance dashboards, making it easier for stakeholders to make informed decisions (Schlueter and Thesseling 2009).

In addition to improving operational energy efficiency, the integration of mathematical optimization, BIM, and lifecycle assessment significantly enhances sustainable building design. This comprehensive framework enables architects and engineers to evaluate multiple design alternatives with respect to energy performance, environmental impact, and material sustainability. By utilizing BIM-lifecycle assessment synergy, it becomes possible to optimize material selection, construction techniques, and long-term energy use, thereby improving the building's energy performance over its entire lifecycle. This approach also supports evidence-based decision-making and promotes sustainability across the AEC industry (Najjar et al. 2019). Another important factor influencing energy performance is building orientation, which affects solar gain, heat loss, and natural lighting. BIM software tools are employed to simulate and analyze the energy impacts of various orientation options. By comparing energy consumption across different building orientations, designers identify the most energy-efficient configuration. Such analyses show that proper orientation can lead to significant long-term energy savings for small- and large-scale building projects (Abanda and Byers 2016).

Despite these benefits, the use of BIM for energy modeling still faces some challenges. One significant limitation is the difficulty in extracting energy-specific information from comprehensive BIM models. For example, industry foundation classes, a common format used in BIM, often contain overly detailed geometrical representations and extraneous data that are not relevant for energy simulations. These complexities hamper the efficiency of performance analysis. To address this, specialized algorithms are used to extract simplified input models tailored for energy calculations directly from lifecycle assessment data, thereby streamlining the modeling process and improving analytical accuracy (Ladenhauf et al. 2016).

4.4.3 BIM in Construction Cost Estimation

The feasibility stage of an infrastructure project lifecycle is often marked by a high degree of uncertainty, particularly in terms of cost estimation. This is primarily due to the lack of reliable and detailed data early in a project. At this initial stage, two of the most significant cost drivers, land acquisition and the construction process itself, are assessed without the availability of fully developed design details. Therefore, improving the accuracy of construction cost estimation during the conceptual design phase is critical, especially as decisions about land procurement must typically be made at this point. Conventionally, cost estimation is done at a stage, often when the conceptual design has been substantially developed or even completed. By this time, it may be too late for stakeholders to make informed and impactful decisions about project scope, design alternatives, or financial viability. Introducing value engineering and cost estimation early in the design stage led to more cost-effective project delivery, enhanced building quality, and greater control and predictability throughout the lifecycle of a project. BIM also facilitates value engineering, allowing project teams to compare alternative materials, construction methods, and design configurations to identify cost-saving opportunities without compromising quality. Through the integration of BIM and value engineering principles, it becomes easier to extract design and construction data, modify designs for cost efficiency, and optimize functionality and performance (Li et al. 2021).

One of the key applications of BIM is automated quantity takeoffs, where BIM models are used to accurately calculate the quantities of different items of work and quantities of materials required, such as steel, concrete, and formwork. These material quantities are linked to the databases containing unit costs, enabling a fast and accurate cost estimation process that minimizes human error (Muratova and Ptukhina 2019). One of the challenges in automated cost estimation is the systematic acquisition of construction-specific information, including construction methods, processes, and equipment needs. To overcome this, methodologies are available to extract and structure such data for seamless integration into BIM-based cost estimation workflows. By analyzing the specifications required for cost estimation, essential construction information is identified and integrated into the BIM model (Bui et al. 2016).

An essential aspect of cost-informed decision-making involves evaluating various design alternatives from a cost perspective. A BIM-based cost estimation system allows stakeholders, particularly project owners, to assess the financial implications of different design scenarios using a dynamic and accurate approach. Any modifications made to components or materials within the BIM model are automatically reflected in the updated cost estimates, thus enabling real-time cost tracking. Additionally, BIM-based platforms are adapted to incorporate national and international cost standards, allowing for standardized and region-specific estimation procedures (Fazeli et al. 2021).

The pre-tender cost estimation, which forms the basis for bidding and tendering, is another critical application of BIM. The estimates must be highly accurate, as these significantly influence project planning and contractor selection. BIM-based cost estimation tools provide a level of precision that surpasses traditional manual methods. Moreover, these tools dramatically reduce the time and effort required for estimating costs, which enhances the overall efficiency of the tendering process (Rashidi et al. 2024).

The evolution from 4D BIM (which links construction sequencing with time) to 5D BIM (which adds cost information) represents a significant advancement in project management. 5D BIM provides a platform for real-time visualization and analysis of cost data, enabling better cost control throughout the construction process. Studies have shown that the variance between planned and actual costs is significantly reduced when 5D BIM is implemented, as opposed to relying on traditional estimation methods (Abdel-Hamid and Abdelhaleem 2023). This capability positions 5D BIM as a transformative approach for enhancing decision-making, increasing transparency, and improving financial oversight in construction projects.

4.4.4 Detection of Design Clashes

During the construction stage of an infrastructure project, frequent revisions to shop drawings are often required due to cognitive and interpretational differences between designers and construction professionals. These two stakeholder groups typically approach clash detection and resolution from distinct perspectives, shaped by their respective backgrounds, domain expertise, and professional experience. While designers focus on architectural intent, spatial constraints, and aesthetic considerations, constructors are more concerned with practical constructability, site logistics, and material installation. These differences lead to misunderstandings and design clashes, which in turn cause delays and inefficiencies in the construction process. To enhance project delivery timelines and reduce rework, it is essential to minimize the number of design model revisions. One effective approach is to utilize a BIM model as a shared reference platform. By making the BIM model accessible to constructors early in the process, potential design conflicts are preemptively identified and evaluated from the construction team's perspective, even before formal design coordination meetings take place. This allows for early-stage clash resolution and informed discussions, leading to more effective compromises and collaborative decision-making. In addition to these procedural improvements, emerging technologies such as artificial intelligence are being adopted to further streamline clash resolution processes. Specifically, artificial intelligence systems that incorporate machine learning algorithms and heuristic optimization techniques have shown promise in automating and enhancing the identification and resolution of design conflicts. These intelligent systems learn from historical clash data, predict likely conflict zones, and propose optimized solutions, thereby improving coordination efficiency and reducing manual effort (Hsu et al. 2020).

4.4.5 *Checking for Regulatory Compliance*

The process of verifying building designs against regulatory requirements and building codes is traditionally a labor-intensive and error-prone task. This manual checking not only consumes considerable time and resources but also introduces a high risk of oversight, especially in complex projects involving multiple disciplines. However, with the advancement and growing adoption of BIM, the automation of code compliance checking has gained significant momentum, offering improved accuracy, efficiency, and consistency. One of the major enablers of this automation is the use of industry foundation classes, which facilitate interoperability and seamless information exchange among various software tools and project stakeholders. Industry foundation classes models serve as a standardized digital representation of building components, which is essential for the development of automated rule-checking systems.

The architecture of an automated code checking system typically comprises several interlinked stages: rule interpretation, building model preparation, rule execution, and rule violation reporting. Among these, the rule interpretation stage is the most critical and complex. This stage involves translating regulatory text, often written in natural language, into a machine-readable and logical format that can be processed by rule-checking algorithms. A range of computational methods and knowledge representation techniques are available to improve the precision and scope of this, ensuring that the encoded rules are both comprehensive and aligned with the regulatory intent, particularly when applied to BIM models expressed in the industry foundation classes format (Ismail et al. 2017).

By integrating rules checking systems into the BIM workflow, compliance with building regulations, such as structural integrity, fire safety, accessibility, and environmental performance, can be monitored automatically and continuously throughout the design process. BIM-based rule-checking tools not only identify potential violations or design shortcomings early but also enable designers to trace the impact of modifications made to address these issues. For instance, when a component in BIM is adjusted to comply with a specific regulation, the system can evaluate how that change may affect other dependent components or potentially introduce new conflicts. This is supported by logic-based rule evaluation interfaces, which help users establish customized rules based on project requirements and map the topological dependencies between different components within the BIM environment (Fan et al. 2019). Despite its potential, the implementation of automated code checking systems faces several challenges. One of the foremost obstacles is the intrinsic complexity and variability of building regulations. The codes/rules often contain ambiguous language, exceptions, and diverse applicability conditions that make them difficult to formalize into executable rules. Furthermore, the sheer volume of regulatory provisions across jurisdictions necessitates a systematic and modular approach to rule formalization in order to make the automation process manageable and scalable (Melzner et al. 2013). However, the integration of BIM with automated rule-checking systems represents a significant advancement in regulatory compliance management. By enabling early detection of design non-conformities, supporting dynamic impact analysis, and reducing manual workload, BIM not only enhances design quality but also contributes to more reliable, efficient, and compliant building projects.

4.5 BIM and GIS Integration in the Design Stage

The energy performance plays a significant role in the design stage of the sustainable lifecycle of an infrastructure. Energy performance is significantly influenced by the surroundings of an infrastructure. However, energy design and its simulation are often done independently without considering the surroundings of an infrastructure. BIM and GIS-integrated web-based system contextualizes low-energy design in an urban energy landscape (Niu et al. 2015). Energy assessment is complex and time-consuming to conduct; the integration of BIM and GIS with a cloud infrastructure facilitates energy assessment for real-time building energy performance detection, visualization, analysis, and optimization across areas that are geographically dispersed (Wu et al. 2014). Preliminary and detailed designs are analyzed in the design stage using BIM and GIS integration (Youn et al. 2014). BIM relies on the local coordinate system that provides no information about an infrastructure with respect to its actual geographic location. BIM and GIS integration allows the transformation of the local coordinates of BIM into its real-world geographic coordinates which is quite helpful in the design stage (Diakite and Zlatanova 2020).

BIM-based practices in transportation infrastructure design are a shift from the traditional design approach towards a highly integrated model, capable of including various design phases along with economic, operational, and environmental concerns. The benefits of the integration between geospatially-referenced data in GIS and BIM models for a more aware design approach are being explored. It minimizes or removes the possible conflicts that arise between the infrastructure design and environmental constraints. BIM and GIS together provide an integrated technical/environmental overview of the design. It allows immediate verification of a project for compliance with the sustainability requirements (D'Amico et al. 2020).

An efficient and sustainable infrastructure is always needed. The need is further gaining momentum as urbanization builds the challenges for sustainable designing, constructing, and operating the built environment by the use of integration of BIM and GIS. The integration provides insight into the information required during the design stage of an infrastructure project (Floros et al. 2020). For GIS, BIM is a promising source of 3D modeling that is used in the study of green buildings and smart cities, etc. (Zhu et al. 2021).

Highway infrastructure plays an important role in ensuring the proper functioning of the nation's transportation. Highway alignment is an essential part of the highway infrastructure design which has significant effects on the surroundings. Identifying optimal highway routes using traditional methods requires significant time, cost, and effort since it requires a comprehensive assessment of multiple factors such as topographical, budgetary, and environmental. The integration of BIM and GIS with the genetic algorithm facilitates in deciding highway alignment in the context of a larger landscape. To support this integration, semantic web-based technologies are used to develop information databases at the semantic level. The visualization of a highway project along with its surroundings provides a comprehensive virtual environment, where a project is effectively planned and designed to have a sustainable solution. This helps in reducing design errors and miscommunications which in turn reduces project risks (Zhao et al. 2019).

BIM and GIS integration is used for estimating land acquisition costs, construction costs, and operation and maintenance costs of road construction projects (Park et al. 2014). Cost estimates to maintain and upgrade road assets are also crucial during budget planning. It requires information regarding the conditions of existing road assets and the cost to improve their level of service which are maintained in GIS.

4.6 Conclusions

In the design stage, architectural and structural designs of an infrastructure are finalized and all the preparations required to begin its construction are made. The work done at this stage is crucial for deciding the right materials and equipment to be used during construction. Environmental concerns sometimes delay a project or prevent it from starting entirely. The completion of the design stage ensures that all the design procedures, design calculations, and documentation in the form of drawings have been undertaken and design approvals have been taken from the relevant approving authorities. A detailed set of construction drawings is also prepared, indicating site layouts, levels, and setting out details of the infrastructure involved. It also includes a detailed schedule of quantities, tender procedures, conditions of contract, and overall construction cost estimate in which GIS is used. Design omissions and errors are very costly both once a contractor is appointed to undertake the construction and at the time of completion of a project. A significant number of variations increase the overall cost of a project sometimes beyond the allocated budget. Although the applications of GIS in the design stage are limited it requires future research.

BIM is transforming the AEC industry by providing data-rich 3D models that enhance collaboration, accuracy, and sustainability throughout the design stage. BIM brings structural analysis, automated rules checking, energy modeling, cost estimation, clash detection, and regulatory compliance into a single digital platform. BIM supports smart buildings by connecting with the Internet of Things and artificial intelligence. BIM also streamlines sustainability assessments, facilitates certification processes, and promotes public engagement in design finalization. Through 5D modeling, performance simulations, and lifecycle analysis, BIM reduces design errors, enhances decision-making, and drives sustainable, cost-efficient, and resilient infrastructure design. BIM-GIS integration enhances sustainable infrastructure design by linking detailed building models with their geospatial context. It supports energy performance analysis, compliance with sustainability requirements, and highway alignment optimization. This integration reduces design errors, project risks, and environmental conflicts while enabling smarter planning of green buildings and urban infrastructure.

4.7 Exercises

Question 4.1: How does GIS help in minimizing flood risks and failures during the design stage of infrastructure projects?

Question 4.2: Why is public participation considered important in the design stage, and how is GIS used to facilitate it?

Question 4.3: Explain how GIS is used in deciding sustainable road alignment in built-up areas.

Question 4.4: What role does GIS play in construction cost estimation, particularly in generating quantity takeoffs and BOQ?

Question 4.5: How does GIS contribute to the visualization of sunlight and deciding the orientation of a building during the design stage?

Question 4.6: In what ways is GIS used for solar energy estimation in buildings and urban planning?

Question 4.7: How does BIM improve collaboration and data exchange among stakeholders during the design stage?

Question 4.8: Discuss how BIM integration with artificial intelligence enhances infrastructure design and decision-making.

Question 4.9: What is the role of BIM in achieving sustainability certifications such as LEED or BREEAM during the design stage?

Question 4.10: Explain how BIM integration with finite element analysis improves structural analysis of buildings.

Question 4.11: What are the benefits of combining BIM with augmented reality in construction design workflows?

Question 4.12: In what ways does BIM enhance sustainable building design and early-stage decision-making?

Question 4.13: How does BIM support environmental impact assessments during the design stage of infrastructure projects?

Question 4.14: How does BIM-based clash detection reduce design errors and improve collaboration between designers and contractors?

Question 4.15: Explain the role of BIM in automated regulatory compliance checking and its challenges.

Question 4.16: What are the benefits of integrating BIM with GIS in the design stage of infrastructure projects?

References

Abanda, F. H., and Byers, L. (2016). "An investigation of the impact of building orientation on energy consumption in a domestic building using emerging BIM (building information modelling)." *Energy*, 97, 517–527.

Abdel-Hamid, M., and Abdelhaleem, H. M. (2023). "Project cost control using five dimensions building information modelling." *International Journal of Construction Management*, 23(3), 405–409.

Al-Kodmany, K. (1999). "Using visualization techniques for enhancing public participation in planning and design: Process, implementation, and evaluation." *Landscape and Urban Planning*, 45(1), 37–45.

Anil, E. B., Akinci, B., Kurc, O., and Garrett, J. H. (2016). "Building-information-modeling–based earthquake damage assessment for reinforced concrete walls." *Journal of Computing in Civil Engineering*, 30(4), 04015076.

Azhar, S., Carlton, W. A., Olsen, D., and Ahmad, I. (2011). "Building information modeling for sustainable design and leed® rating analysis." *Automation in Construction*, 20(2), 217–224.

Bansal, V. K., and Pal, M. (2007). "Potential of geographic information systems in building cost estimation and visualisation." *Automation in Construction*, 16(3), 311–322.

Bansal, V. K., and Pal, M. (2009). "Extended GIS for construction engineering by adding direct sunlight visualisations on buildings." *Construction Innovation*, 9(4), 406–419.

Bhatarai, R., Banihashemi, S., Shakouri, M., and Antwi-Afari, M. (2025). "Integration of augmented reality with building information modeling: design optimization and construction rework reduction perspective." *Archives of Computational Methods in Engineering*, 32, 2233–2254.

Bui, N., Merschbrock, C., and Munkvold, B. E. (2016). "A review of Building Information Modelling for construction in developing countries." *Procedia Engineering*, 164, 487–494.

Chang, Y. T., and Hsieh, S. H. (2020). "A review of Building Information Modeling research for green building design through building performance analysis." *Journal of Information Technology in Construction*, 25, 1–40.

Cheng, M. Y., and Yang, C. Y. (2001). "GIS-based cost estimate integrated with material layout planning." *Journal of Construction Engineering and Management*, 127(4), 291–299.

Chi, H. L., Wang, X., and Jiao, Y. (2015). "BIM-enabled structural design: impacts and future developments in structural modelling, analysis and optimisation processes." *Archives of Computational Methods in Engineering*, 22, 135–151.

Choi, J., Lee, S., and Kim, I. (2020). "Development of quality control requirements for improving the quality of architectural design based on BIM." *Applied Sciences*, 10(20), 7074.

D'Amico, F., Calvi, A., Schiattarella, E., Di Prete, M., and Veraldi, V. (2020). "BIM and GIS data integration: a novel approach of technical/environmental decision-making process in transport infrastructure design." *Transportation Research Procedia*, 45, 803–810.

de Lima Montenegro, J. G. C., Zemero, B. R., de Souza, A. C. D. B., de Lima Tostes, M. E., and Bezerra, U. H. (2021). "Building Information Modeling approach to optimize energy efficiency in educational buildings." *Journal of Building Engineering*, 43, 102587.

Diakite, A. A., and Zlatanova, S. (2020). "Automatic geo-referencing of BIM in GIS environments using building footprints." *Computers, Environment and Urban Systems*, 80, 101453.

Fan, S. L., Chi, H. L., and Pan, P. Q. (2019). "Rule checking Interface development between building information model and end user." *Automation in Construction*, 105, 102842.

Fazeli, A., Dashti, M. S., Jalaei, F., and Khanzadi, M. (2021). "An integrated BIM-based approach for cost estimation in construction projects." *Engineering, Construction and Architectural Management*, 28(9), 2828–2854.

Floros, G. S., Ruff, P., and Ellul, C. (2020). "Impact of information management during design & construction on downstream BIM-GIS interoperability for rail infrastructure." *ISPRS Annals of the Photogrammetry, Remote Sensing and Spatial Information Sciences*, 6(4), 61–68.

Gourlis, G., and Kovacic, I. (2017). "Building Information Modelling for analysis of energy efficient industrial buildings–A case study." *Renewable and Sustainable Energy Reviews*, 68, 953–963.

Grover, R., Li, P., and Froese, T. M. (2015). "The interface between building information models and the public." *International Construction Specialty Conference of the Canadian Society for Civil Engineering (ICSC)*.

Hsu, H. C., Chang, S., Chen, C. C., and Wu, I. C. (2020). "Knowledge-based system for resolving design clashes in building information models." *Automation in Construction*, 110, 103001.

Inyim, P., Rivera, J., and Zhu, Y. (2015). "Integration of building information modeling and economic and environmental impact analysis to support sustainable building design." *Journal of Management in Engineering*, 31(1), A4014002.

Ismail, A. S., Ali, K. N., and Iahad, N. A. (2017). "A review on BIM-based automated code compliance checking system." *2017 International Conference on Research and Innovation in Information Systems (ICRIIS)*,1-6, IEEE.

Jiang, F., Ma, L., Broyd, T., Chen, K., Luo, H., and Pei, Y. (2022). "Sustainable road alignment planning in the built environment based on the MCDM-GIS method." *Sustainable Cities and Society*, 87, 104246.

Kassem, M., Iqbal, N., Kelly, G., Lockley, S., and Dawood, N. (2014). "Building information modelling: protocols for collaborative design processes." *Journal of Information Technology in Construction*, 19, 126–149.

Koo, H. J., and O'Connor, J. T. (2022). "Building information modeling as a tool for prevention of design defects." *Construction Innovation*, 22(4), 870–890.

Ladenhauf, D., Berndt, R., Krispel, U., Eggeling, E., Ullrich, T., Battisti, K., and Gratzl-Michlmair, M. (2016). "Geometry simplification according to semantic constraints: Enabling energy analysis based on building information models." *Computer Science-Research and Development*, 31, 119–125.

Li, X., Wang, C., and Alashwal, A. (2021). "Case study on BIM and value engineering integration for construction cost control." *Advances in Civil Engineering*, 2021(1), 8849303.

Liu, S., Meng, X., and Tam, C. (2015). "Building information modeling based building design optimization for sustainability." *Energy and Buildings*, 105, 139–153.

Love, P. E., Edwards, D. J., Han, S., and Goh, Y. M. (2011). "Design error reduction: toward the effective utilization of building information modeling." *Research in Engineering Design*, 22, 173–187.

Mardaljevic, J., and Rylatt, M. (2003). "Irradiation mapping of complex urban environments: an image-based approach." *Energy and Buildings*, 35(1), 27–35.

Melzner, J., Zhang, S., Teizer, J., and Bargstädt, H. J. (2013). "A case study on automated safety compliance checking to assist fall protection design and planning in building information models." *Construction Management and Economics*, 31(6), 661–674.

Migliaccio, G. C., Zandbergen, P., and Martinez, A. A. (2009). "Assessment of methods for adjusting construction cost estimates by geographical location." Proceeding of International Conference of ASCE on *Construction Research Congress*, 886–895.

Migliaccio, G. C., Zandbergen, P., and Martinez, A. A. (2015). "Empirical comparison of methods for estimating location cost adjustments factors." *Journal of Management in Engineering*, 31(2), 04014037.

Muratova, A., and Ptukhina, I. (2019). "BIM as an instrument of a conceptual project cost estimation." In *International Scientific Conference on Energy, Environmental and Construction Engineering*, 469–477, Cham: Springer International Publishing.

Najjar, M., Figueiredo, K., Hammad, A. W., and Haddad, A. (2019). "Integrated optimization with building information modeling and life cycle assessment for generating energy efficient buildings." *Applied Energy*, 250, 1366–1382.

Niu, S., Pan, W., and Zhao, Y. (2015). "A BIM-GIS integrated web-based visualization system for low energy building design." *Procedia Engineering*, 121, 2184–2192.

OpenAI (2025). ChatGPT (Version 5.1) [Large language model]. https://chat.openai.com/

Park, T., Kang, T., Lee, Y., and Seo, K. (2014). "Project cost estimation of national road in preliminary feasibility stage using BIM/GIS platform." Proceeding of International Conference of ASCE on *Computing in Civil and Building Engineering-2014*, 423–430.

Quan, S. J., Li, Q., Augenbroe, G., Brown, J., and Yang, P. P. J. (2015). "A GIS-based energy balance modeling system for urban solar buildings." *Energy Procedia*, 75, 2946–2952.

Rashidi, A., Chan, D. W., Ravanshadnia, M., Sarvari, H., and Tajaddini, A. (2024). "Applying building information modelling (BIM) technology in pre-tender cost estimation of construction projects: a case study in Iran." *Buildings*, 14(5), 1260.

Ratti, C., Baker, N., and Steemers, K. (2005). "Energy consumption and urban texture." *Energy and Buildings*, 37(7), 762–776.

Ren, X., Fan, W., Li, J., and Chen, J. (2019). "Building Information Model–based finite element analysis of high-rise building community subjected to extreme earthquakes." *Advances in Structural Engineering*, 22(4), 971–981.

Rylatt, M., Gadsden, S., and Lomas, K. (2001). "GIS-based decision support for solar energy planning in urban environments." *Computers, Environment and Urban Systems*, 25(6), 579–603.

Schlueter, A., and Thesseling, F. (2009). "Building information model based energy/exergy performance assessment in early design stages.|" *Automation in Construction*, 18(2), 153–163.

Shrestha, K., and Shrestha, P. (2014). "A GIS-enabled cost estimation tool for road upgrade and maintenance to assist road asset management systems." Proceeding of International Conference of ASCE on *Construction Research Congress*, 1239–1248.

van Eldik, M. A., Vahdatikhaki, F., dos Santos, J. M. O., Visser, M., and Doree, A. (2020). "BIM-based environmental impact assessment for infrastructure design projects." *Automation in Construction*, 120, 103379.

Wong, K. D., and Fan, Q. (2013). "Building information modelling (BIM) for sustainable building design." *Facilities*, 31(3/4), 138–157.

Wu, W., Yang, X., and Fan, Q. (2014). "GIS-BIM based virtual facility energy assessment (VFEA) Framework development and use case of California State University, Fresno." Proceedings of *Computing in Civil and Building Engineering*, 339–346.

Xu, X., Mumford, T., and Zou, P. X. (2021). "Life-cycle building information modelling (BIM) engaged framework for improving building energy performance." *Energy and Buildings*, 231, 110496.

Youn, J., Kang, T. W., Choi, H. S., and Kim, C. (2014). "Derivation of BIM/GIS platform application scenarios and definition of specific functions in construction planning and design." *Journal of the Korea Academia-Industrial Cooperation Society*, 15(12), 7340–7349.

Zhao, L., Liu, Z., and Mbachu, J. (2019). "Highway alignment optimization: an integrated BIM and GIS approach." *ISPRS International Journal of Geo-Information*, 8(4), 172.

Zhong, D., Li, J., Zhu, H., and Song, L. (2004). "Geographic information system based visual simulation methodology and its application in concrete dam construction processes." *Journal of Construction Engineering and Management*, 130(5), 742–750.

Zhu, J., Tan, Y., Wang, X., and Wu, P. (2021). "BIM/GIS integration for web GIS-based bridge management." *Annals of GIS*, 27(1), 99–109.

Chapter 5

The Construction Planning Stage

5.1 Learning Objectives

After the completion of this chapter, readers will be able to:

- understand the need and use of GIS in the construction planning stage,
- understand the use of GIS as an alternative for scheduling infrastructure construction projects,
- explore the use of GIS in space conflict analysis and planning for safety, and
- understand the application areas of BIM and the integration of BIM and GIS in the construction planning stage.

5.2 Introduction

Planning is a bridge between ideas and their execution. It is a fundamental requirement of infrastructure construction projects but a challenging task. It is quite complex and highly individualistic in nature. It involves the development of a work breakdown structure, choice of construction methods/technologies, estimation of the time duration of individual activity, identification of constraints among different activities, developing precedence relationships among activities, deciding the time duration of a project, and finally, development of an execution plan. A planning party investigates site conditions/constraints, site topography, contract documents, two-dimensional (2D) drawings, and three-dimensional (3D) models of infrastructure to develop an executable plan. A large amount of such information and interdependency among different activities impose a heavy burden on a planning party in order to develop an executable construction plan.

Keeping the complexities of the planning process in view, information technologies are being used in the planning stage of infrastructure projects. Information technology-based tools like knowledge-based systems, computer graphics, and a combination of the two have been reported in the literature for planning the construction of infrastructure projects. The first main research direction in the area of planning was focused on developing knowledge-based systems for automating

DOI: 10.1201/9781003759348-5

the planning process. The construction knowledge gained from the experience is coded in the form of rules, regulations, and procedures to generate an execution plan. The knowledge-based systems mainly depend upon the programmed knowledge and are heuristic in decision-making. A major problem in such systems is that they reduce or eliminate the role of a planning party. A planning party is mainly involved in the data input and has minimum interaction and control in the planning process.

The second main research direction utilized computer graphics (3D modeling, four-dimensional (4D) tools, computer-based simulations, etc.) to develop animations of a 3D model over the time duration of a project to facilitate the planning process (Koo and Fischer 2000). Virtual reality-based systems supplement 4D tools by adding temporary works and corresponding activities to set and dismantle them. These systems help a planning party in reviewing a construction plan by identifying missing activities, logical errors in the schedule, spatial feasibility, identifying time-space conflicts, constructability analysis, finding alternate construction methods, undertaking a virtual pre-construction rehearsal, etc.

The knowledge-based systems and computer graphics were integrated in the third main research area. The programmed knowledge was combined with computer graphics to automate the infrastructure construction planning process. The system automatically generates construction activities, their time durations, and execution schedules after entering the required information such as drawings, construction methods, available resources, etc. Recent developments in construction planning utilize building information modeling (BIM) that provides strong premises to overcome the fragmented nature of the planning process (Isikdag et al. 2008).

The construction plan of an infrastructure must not be developed in isolation, that is, without considering its immediate surroundings. It is affected by or has an impact on existing facilities/utilities in the immediate surroundings. The computer-aided design (CAD) and BIM-based planning tools used in the AEC industry primarily consider the internal geometry of an infrastructure to be developed. The effects of surroundings and nearby existing facilities/utilities on the construction plan are considered manually by planners based on their judgments and experience. Although, CAD-based planning tools help planners in developing an execution schedule, cost estimate, and 4D simulation but lack in considering surroundings, which include topography, spatial constraints, locational factors, etc. The use of GIS is reported to consider surroundings, existing facilities/utilities, and to carry out various geospatial analyses; hence, GIS facilitates planners in developing an executable construction plan (Bansal 2014). Further, the AEC industry faces challenges because of different information formats used to represent graphical and non-graphical information, which are difficult to bridge together. This is called the interoperability problem in the AEC industry. Lack of interoperability is a major obstacle that limits workflow usability due to difficulty in using information from building geometries, coordinates, location, thermal zones, etc. The use of BIM is reported to minimize the interoperability problem in the AEC industry. However, in BIM an infrastructure is still planned in isolation. This chapter in the following sections, covers the application areas of GIS, BIM, and integration of BIM and GIS in the planning stage of infrastructure projects, taking their surroundings into consideration.

5.3 Construction Planning

Infrastructure construction planning is divided into *pre-project planning*, *pre-bid planning*, *pre-construction planning*, and *during-construction planning*. *Pre-project planning* is an owner-driven planning process of gathering sufficient information about a project on the potential risks versus probable success so that an owner can decide whether a project should proceed or not. The *pre-bid planning* is performed by contractors before the submission of their bids to take part in the bidding process. It is a contractor-driven planning process. *Pre-construction planning* begins with the award of a project and ends at the point of time when construction starts. The *pre-construction planning* is of vital importance for the successful execution and delivery of a project. It is a macro-level planning done by a contractor who gets the construction project. It is mainly concerned with design review, finalization of execution sequence, constructability analysis, site planning for major operations, logistics planning, deciding major equipment required, etc. It minimizes risks, material wastage, overheads, and maximizes productivity during the construction. The *pre-construction planning* is essential to increase productivity, decrease construction costs, improve efficiency, and increase profitability. D*uring-construction planning* is contractor-driven planning done by a contractor during construction. In *during-construction planning*, decisions are made at the micro-level to assist the field crew during the execution of day-to-day operations on a site. D*uring-construction planning* is not a one-time activity; it continues throughout the construction stage of an infrastructure project.

5.4 Need for GIS in the Planning Stage

The construction planning of an infrastructure project is often considered a bridge between the structural design and execution. The development of an executable construction plan is quite complex and highly individualistic in nature. The complexities and uncertainties in an infrastructure construction project make its planning difficult. Hence, the development of an executable construction plan is always a critical task in front of planners. Any construction, either big or small, becomes an important activity in a locality for a specific time period. Therefore, construction should not be planned in isolation as a single entity but in close relation with respect to all other existing facilities/utilities, neighborhoods, topography, etc. (Underwood and Isikdag 2010). A construction plan must consider the ways by which surroundings influence it as well as how it is influenced by social, economic, and environmental factors in its surroundings. Further, construction planning decisions impact surroundings, specifically in the case of sustainable development (Hammad et al. 1993).

CAD or BIM mainly focuses on the geometric information of an infrastructure; on the other hand, GIS considers its surroundings (Isikdag et al. 2008). Hence, only the use of CAD or BIM in the construction planning stage indicates that infrastructure is being planned in isolation, that is, without considering its surroundings. CAD or BIM lacks in the evaluation of a construction site with respect to flooding, finding changes in the drainage pattern due to the alteration in topography, route planning for vehicles carrying consignments from different access routes to a construction site, etc. A construction plan is linked to the nearby existing facilities/utilities in the surroundings of an infrastructure that provides water, energy, resources, communications, etc. The layouts of existing services like: electric supply lines, water distribution systems, sewerage networks, etc., which play a major role in any construction activity are generally stored in the GIS environment (Gunes and Kovel 2000). The locational aspects/layouts of existing facilities/utilities are also important in construction planning along with the geometric and semantic information of an infrastructure to be developed (Wong and Jusuf 2007). Further, any construction, specifically in hilly regions where topography plays a major role, should not be planned without a tool having geo-spatial analysis capabilities which are available in GIS. Hence, a 3D model of an infrastructure is prepared along with its surroundings and topography (Bansal and Pal 2008). However, CAD or BIM ignores all such collaborations but mainly focuses on the geometric information of an infrastructure only. The database management capabilities of GIS also provide the mechanism for querying and retrieving construction-specific information to support construction planning.

Infrastructure construction planning requires visualization of configurations, characteristics, and spatial constraints among various components of an infrastructure to be developed. Refined spatial information is generated in GIS by querying planning data. GIS helps in virtually measuring horizontal distances, elevations, directions, slopes, and angles on the ground from the surface model needed in location identification, planning of a site layout, and planning of available workspace. GIS provides information about a location (latitude, longitude, and elevation) in space and distance from a targeted location. GIS-based spatial analyses that support construction planning are briefly discussed below:

Selection of components: GIS helps in the selection of components of an infrastructure satisfying a particular set of constraints, component/s corresponding to an activity, and specifications/cost/resource requirements of an activity from a database. GIS is used to find locations of components corresponding to a particular activity of interest, as shown in Figure 5.1(a) and (b) shows how GIS is used in finding existing facilities within a specified distance surrounding the location under consideration.

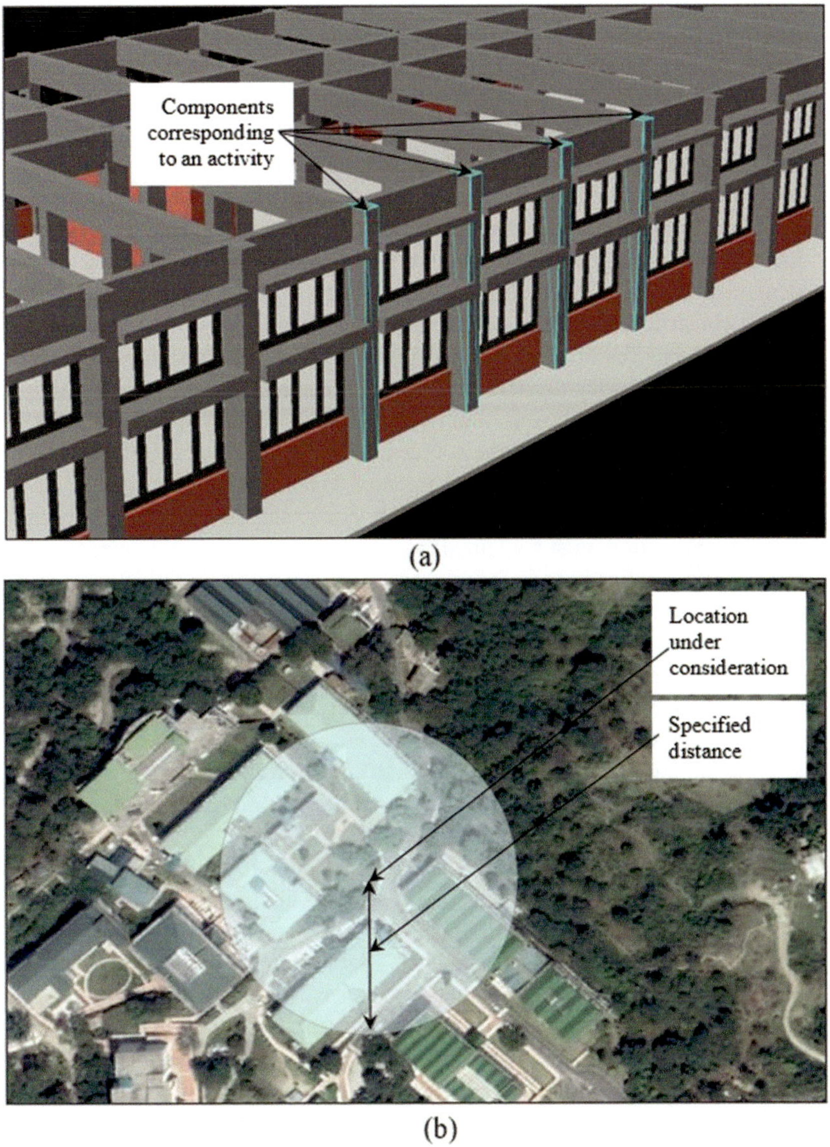

Figure 5.1 (a) Finding components corresponding to an activity and (b) finding existing facilities surrounding the location under consideration

Location-based selection: Selects features/components from a layer, based on their location relative to features/components in another layer. For example, if a planner wants to know how many activities are in the neighborhood of an activity under consideration, a planner may select all features/components that touch the component/feature corresponding to an activity under consideration. Location-based selection may be used to select various components/features in one layer that are near or overlap components/features in the same or another layer. Intersect, contain, and identical to are the common location-based selection tools available in GIS that help in selecting components/features based upon their location relative to other components/features in the construction planning process. Intersect generally integrates two sets of components/features maintained in two different layers, preserving only those components/features that fall within the spatial extent common to both intersecting layers. Intersect returns components/features that geometrically share/s a common part with the source components/features. Contain ensures that each component/feature of the source layer must fall inside the geometry of the targeted components/features. For example, an area, representing a construction site, must contain all working area requirements corresponding to various activities in the schedule. Identical is used to find identical components/features. Features are considered identical if their geometry are strictly equal. This is used to compare the geometry of two components or working areas. Another widely used function is to dissolve features based on their attributes; this removes boundaries or nodes between adjacent features that have the same values for a specified attribute. For example, dissolve may be used to group all components corresponding to an activity based on the name of the activity as an attribute. It also joins non-adjacent components/features to create a single multipart feature (Bansal 2014).

Area-based selection: Highlights all components and corresponding activities that occur within a specified area. Planning queries to determine components occupying a specified area, all activities occurring in that area, and resources that will use that area are performed using area-based selection in GIS. In this, a buffer (or buffers) is created with a size equal to the specified distance around the source components/features, and then it returns all components/features intersecting a buffer (or buffers). It helps in determining the work envelope of equipment, which is an area that equipment can reach based on various constraints during its operation from a location on a site. A construction planner can move and place an envelope of equipment on a site to determine its most feasible location for examining operating clearances. This may also be used to retrieve components within the specified distance from a particular location or a room. For example, GIS is used to find the components of a building that are accessible to a crane when installed on a particular location, as shown in Figure 5.2.

Selection within a time-frame: Highlights project information like all activities that occur within a specified time frame or all resources performing work in that time frame. GIS helps in the selection of building information/activities/

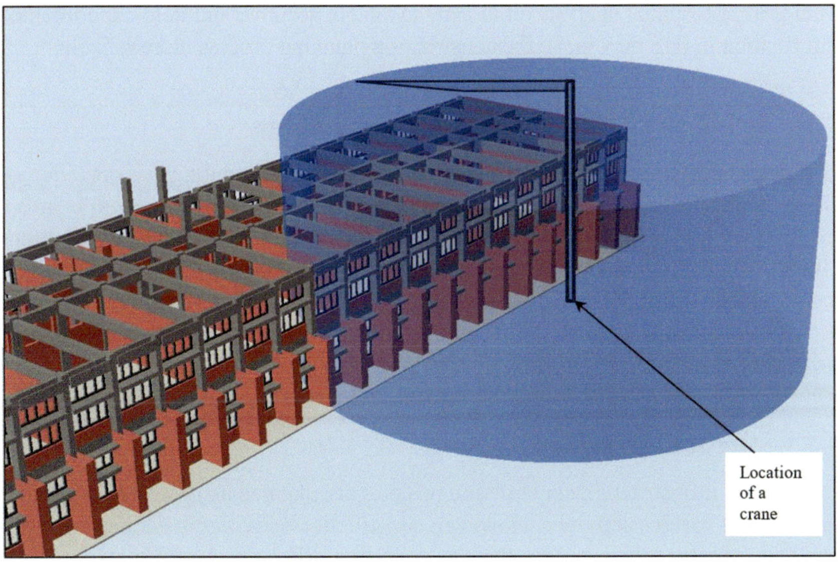

Figure 5.2 Finding components of a building which are approachable to a crane when installed on a particular location

components that follow a set of constraints and then again filters the selection with another set of spatial and non-spatial constraints (Bansal and Pal 2009).

Information database: Infrastructure construction planning involves the interpretation and investigation of project-specific information available in graphical and non-graphical forms obtained from different sources. The large volumes of such information result in complexities in their use in the planning process. Hence, the capabilities of GIS are used in the development of the organized planning database to maintain and use information in time to facilitate the planning process. Refined graphical and non-graphical information is extracted from the planning database by graphical and non-graphical querying. Various graphical operations on graphical information and non-graphical operations on non-graphical information in a GIS environment improve the construction planning process.

Geospatial technologies are also being used in cadaster, land use planning, demolition planning, emergency response planning, navigation and routing, and planning the delivery of goods and services. Different types of site information required in the construction planning process of large infrastructure projects like residential colonies, industries, and hydro-electrical projects are developed in the geospatial environment. The layouts of different types of existing utility systems, like electricity, gas supply, water distribution, and sewage networks, are now GIS-based, which plays a major role in site management and construction planning of large infrastructure projects.

Therefore, geospatial analysis on already available accurate and field-experimented information in GIS may make the construction planning process more reliable.

5.5 GIS in the Construction Planning Stage

GIS improves infrastructure projects' construction scheduling by considering site-specific factors, accurate production rates, efficient resource allocation, safety planning, and visual monitoring of progress in complex terrains. 4D modeling in GIS integrates safety planning with construction scheduling, enabling hazard identification, safe execution sequencing, complex visualizations, and proactive risk management. Different applications of GIS are discussed in the following sections:

5.5.1 Site Specific Production Rates

Most of the infrastructure construction projects are planned using standard production rates of various activities. However, construction activities in congested or hill areas are greatly influenced by the site conditions. The standard production rates of various activities vary with varying site conditions in terms of the difficulty in executing of various activities. Several environmental, locational, and geological factors influence the productivity of various activities in infrastructure projects. Earthwork in landslide-prone areas is extremely difficult, hence, carried out with great caution, leading to a loss in productivity. Hazards-prone areas not only lead to a reduction in productivity but also require rework in case of any hazards or critical site conditions. GIS is used for the development of a zonation map for assigning variable production rates to activities involved in different locations. A zonation map represents a large site into small zones and ranks zones according to the degree of criticality. A zonation map helps in suitably deciding the production rates of various activities as per the site conditions, adding extra activities related to precautionary work, planning for safety measures against any accidents, and including additional time between successive activities as a buffer to develop an executable schedule (Sharma and Bansal 2018).

Zonation maps are important during the planning and construction stage from the viewpoint of the calculation of production rates of various activities, construction safety, and long-term stability after construction. The contributory factors in the development of a zonation map in GIS are slope angle, aspect, rock type/strata, distance from already existing roads, distance of the market from a construction site, etc. A layer corresponding to each factor contributing toward the criticality of a site is developed. Weights are calculated and assigned to each factor in GIS to perform a weighted overlay analysis in GIS. In the weighted overlay analysis, the weight given to a factor and the weight given to a class of that factor are multiplied together and the value is assigned to a grid cell. Each grid cell has a unique value. When the layers corresponding to factors are overlaid over each other, the values in the corresponding grid cells are added.

Construction sites having steep slopes, geologically unstable strata, proximity to streams, and congested areas contribute towards criticality. These sites have lower production rates of construction activities due to extra precautions, mitigation measures, and sometimes rework required. Weighted overlay analysis divides a site into small zones depending on the criticality. The zonation map classifies a large site into nil, low, medium, high, and very high criticality zones. The zonation map is used for modifying standard production rate values on each location of a project (Sharma et al. 2014).

Location-based planning (LBP) is the most suitable method for planning linear repetitive construction projects like highways, railways, pipelines, and canal construction. The linear construction in plain areas is planned using uniform task production rates throughout the project length. However, topographic variations in a hilly terrain have a huge impact on the task production rates. Hence, to prepare an accurate and reliable execution plan, location-specific geographic information is used to arrive at more realistic task production rates (Sharma et al. 2016).

5.5.2 Scheduling of Non-Repetitive Construction Projects

The AEC industry uses bar charts, critical path method (CPM), or line of balance (LOB) techniques to develop an executable schedule of infrastructure construction projects. Different activities in the execution schedule are related to one or more components of an infrastructure project under consideration. To visualize the execution sequence of an infrastructure project, planners mentally associate the components of the drawings with the related activities present in the schedule. There is no dynamic linkage between the construction schedule and the components of drawings in the commercially available planning and scheduling tools (Koo and Fischer 2000). The interpretation of a schedule without any link between its activities and graphical information is difficult because an infrastructure project may contain hundreds of activities. This makes the project execution schedule difficult to check for its completeness. Also, the interpretation of a plan varies with the experience of the persons involved due to the difficulty in mentally linking each component of drawings with the corresponding activities of the schedule (Poku and Arditi 2006).

The lack of linkage between the execution schedule and the components of a drawing forced the researchers to link the schedule with the 3D model of an infrastructure which led to the development of a 4D model. It allows a planner to virtually visualize the project execution process in a way as it would actually be built. 4D modeling increases the comprehensibility of the project schedule and allows users to detect scheduling conflicts before the project execution. The use of 4D modeling helps in allocating resources more effectively, avoiding schedule conflicts, examining constraints, evaluating alternative construction methods, etc. (Akinci et al. 2002).

5.5.2.1 Critical Path Method in GIS

In order to develop a 4D model of the execution sequence, a construction schedule is developed by using widely used scheduling tools. A 3D model of an infrastructure is developed by using CAD or BIM-based software tools and linked with the schedule in some other software tool. Linking a 3D model developed in CAD or BIM-based software tools with the schedule is a tedious and time-consuming process. As the infrastructure construction progresses, the schedule is updated frequently and scheduling computations are carried out many times on the updated information. Therefore, in the schedule evaluation and updating process, the 4D model of the execution process needs to be disconnected and connected depending upon the frequency of updating, so that the required updating/correction may be carried out in suitable project management tools. For example, the schedule is updated in scheduling tools, CAD or BIM-based software tools are used to update the components, and for linking the schedule and components, other software tools are utilized. This problem was resolved by developing the schedule and linking the activities in the schedule with the corresponding components in a single GIS environment (Bansal and Pal 2008).

The schedule is the timetable for the execution of various activities of an infrastructure project. It involves the identification of all possible activities of a project and defining the precedence relationship among them to arrange all in an execution sequence. The time duration for each activity is estimated. GIS is used to handle the CPM basic scheduling computations and link the developed CPM-based schedule with the 3D model in the same environment to develop the 4D model of the execution sequence (Bansal and Pal 2011). A CPM-based schedule in GIS is developed in the form of a table. The output table has a fixed number of columns; the number of rows depends on the number of activities in the schedule. A user inputs the information about the activity description, from node (predecessor event number of an activity), to node (successor event number), and activity time duration. The connection between an activity and corresponding 3D components is established through a key field in the schedule and attribute tables of various components. The GIS-based scheduling tool calculates the time duration of a project and the critical path. The output table provides values of activity times (earliest start time, earliest finish time, latest start time, and latest finish time), various types of floats (total float, free float, interfering float, and independent float), and criticality of network activities.

The CPM tool in GIS also provides a schedule in the form of a single calendar date such as 7 working days in a week, 6 working days in a week, and 5 working days in a week. The output of the calendar date scheduling is activity times in the form of day and date, for example, Friday, 14 May 2024. The schedule may be displayed in the form of a bar chart. In the bar chart, activities are listed in order of construction priorities on the y-axis, while the time duration is plotted on the x-axis. The main advantage of a bar chart in GIS is that when a bar on the bar chart is clicked, GIS displays the scheduling information related to that particular activity on the computer screen. A bar chart in GIS is dynamically linked with the execution schedule so as to reflect the updated information (Bansal and Pal 2008).

Scheduling in GIS also facilitates schedule updating. Updating accepts the changes in the estimated time durations of activities already available in the schedule and the addition and deletion of an activity. In order to update the schedule, entire activities do not need to be input again; only the affected inputs are altered directly in the schedule. Care is taken so that no errors, gaps, or loops are created in the project network while adding or deleting an activity. The scheduling program in GIS also allows adding different types of resources required for each activity and generates a resource profile for each type of resource required. The database management capabilities of GIS are utilized to maintain and update the construction resource database to facilitate project planning. The non-spatial information (like materials type and quantities, labor requirements, safety and quality control recommendations, etc.) corresponding to different activities in the schedule can be extracted from the database maintained in GIS and can be linked with respective activities present in the schedule. Selecting an activity in a CPM-based schedule in GIS automatically highlights the attributes related to it in the resource database.

5.5.2.2 Advantages of Scheduling in GIS

CPM-based scheduling in GIS allows multiple initial and terminal events in the construction project networks. The events of the network, representing the execution sequence, need not to be numbered in ascending order; that is, each successor event number need not to be larger than the predecessor event number. Scheduling in GIS permits random numbering of events, provided no two nodes receive an identical number. Any two events in a network may be directly connected by more than one activity. The requirement of a unique pair of start and end numbers for each activity is not necessary. The project activities may be stored/listed in a variety of useful ways, such as sorting in ascending/descending order on any field in the schedule table. The calculations like sum, count, mean, maximum, minimum, range, variance, and standard deviation of a particular column in an output table is also possible (Bansal and Pal 2011).

5.5.3 Scheduling of Repetitive Construction Projects

The network-based methods like CPM and program evaluation and review technique (PERT) are more commonly used in the planning of non-repetitive construction projects. However, they have limitations when applied to repetitive construction projects as the network becomes relatively large, complicated, and continuous crew engagement is not ensured. Further, there is no information about work locations, progress direction, and task production rates at any point of time during project execution. Literature suggests the use of linear scheduling methods for repetitive infrastructure construction projects (Kenley and Seppanen 2010). The linear scheduling methods do not emphasize much on minimizing the time duration of a project but rather on productivity enhancement by ensuring resource continuity through balanced production rates of various tasks. Linear scheduling methods take less time and effort to develop the schedule, a relatively small number of repetitive tasks are neatly represented on a graph, and are easily read and understood by the field staff. The graphical representation of a linear schedule provides instant information on work location, progress direction, task production rates to be maintained, and helps in ensuring work continuity and detecting space and time conflicts.

LOB is one of the oldest and most widely used linear scheduling methods. It is used to handle projects involving repetitive activities. An activity to be carried out at multiple work locations is treated as a single activity. This reduces the total number of activities to be presented on the execution schedule. For example, a hundred houses in a multiple housing scheme consisting of 25 activities to be carried out in each house, resulting in a total of 2500 activities, is scheduled using just 25 distinct activities in LOB. This simplification offers some major advantages over the bar charts and network-based methods: (i) it conveys important production information like work locations, progress direction, and production rates in easy to visual graphical format, (ii) it helps in detecting space and time conflicts among various activities, and (iii) it takes less time and effort to develop than network-based methods. Its major applications are found in multiple housing schemes, high-rise buildings, pipelines, railways, and highways (Sharma et al. 2014). GIS-based geographical visualization is used for planning, scheduling, and monitoring the construction which is repetitive in nature (Seo and Kang 2006; Sonmez and Uysal 2008). The visualization of geographical conditions and their impact on the progress rates of various activities are easily explored by using time-lapse simulation in GIS.

5.5.4 *Location-Based Planning*

The typical characteristics of projects that use network-based scheduling methods are: unique work location, project site condition, unique components, highly sequential activities, and focus on resource optimization rather than resource continuity (Kenley and Seppanen 2010). These characteristics do not match with those of construction projects where there are continuous locations and activities, a large amount of onsite fabrication is involved, simultaneous activities are frequently carried out, and resource continuity is more important than resource optimization. Hence, network-based techniques are not suitable for planning and scheduling of such projects in general. LBP is more logical and widely used to plan and schedule such projects.

LBP shifts focus from activities to locations. Instead of breaking down a project work into activities, a project is broken down into locations, forming a location breakdown structure (LBS) of a repetitive project. Each location in LBS contains project data limited to its scope of work. The quantities of work within each location are computed and a task is defined for each quantity of work. The crews assigned to each task move from one location to the next until they finish all locations. A location is marked complete when all tasks in that location are completed. The LBP process optimizes production through work continuity for resources, improved constructability, and reduced risk in project delivery and overall production cost. This ensures a continuous flow of crews from one location to another without any idle time.

Once the LBS is defined and locations are formed, the quantities of work within each location are estimated. The quantity of work for a given task in a given location is called location-based quantity (LBQ). It includes all the work that must be completed by a crew before a location is marked complete for that task and the crew is allowed to move to the next location. LBQ are used in both scheduling and cost estimating. A planner ensures that all quantities defined for a location are assigned tasks so that nothing is left out of the execution schedule. A schematic representation of LBP is shown in Figure 5.3.

LBP defines a task as a sequence of activities carried out by a single crew to achieve a common goal. For example, the task "construction of a pavement layer" includes activities to be performed in sequence: (i) blowing dust out from the underlying layer, (ii) sprinkling water, (iii) spreading aggregates, (iv) spraying bitumen, and (v) rolling of the layer. Each location in the LBS contained its own independent work quantities of various construction elements and hence had its own independent location-based tasks (LBT). A task required to be performed for LBQ within a location is called LBT. LBQ and associated LBT may vary in each location depending on the construction elements that lie within that location. Once the LBQ and LBT are defined, the computation of time durations and costs of all LBTs is performed. The duration of each task within a location is estimated from the LBQ and location-specific task productivity rate. The tasks are arranged in a logical sequence and the start and end dates of all tasks in all locations are computed.

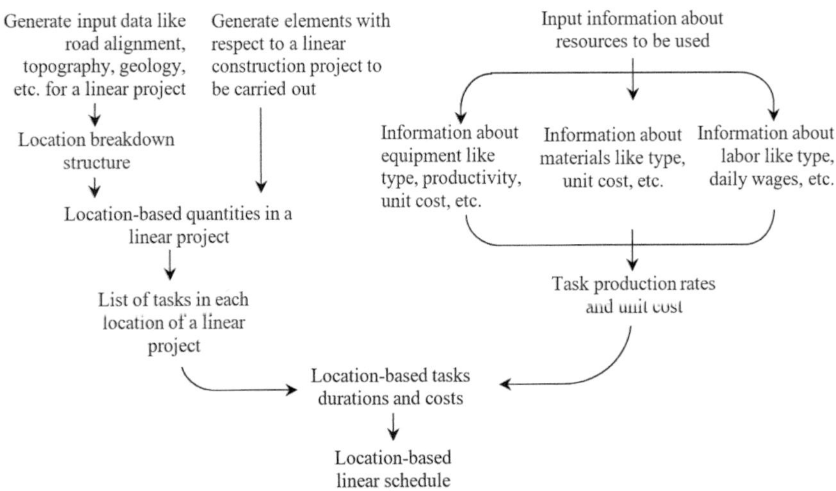

Figure 5.3 A schematic representation of location-based planning

GIS has three main features that make it a useful tool for LBP and location-based scheduling (LBS) of linear infrastructure construction projects: cartographic capabilities for representing geographic information, a relational database management system for maintaining project information, and spatial and non-spatial analysis capabilities. The information databases are maintained in a single GIS environment where their updating is efficient and accurate. Also, these are utilized simultaneously for any integrated query or analysis.

GIS brings information from different sources in different formats to one place by storing them with reference to their geographic location. The visualization of information in one place has the potential to improve planning decisions by considering them all in totality. Like BIM for building construction planning, GIS has the potential to become a standalone tool for highway construction planning. GIS is used in various stages of highway construction planning and has found applications in information management, optimum earthwork planning, project monitoring and control, cost estimation, schedule visualization, etc. Topography modeling in GIS helps in visualizing the alignment of highway infrastructure projects along with their surroundings and also in estimating and planning earthwork, which is the most complicated and tedious task to plan in a highway construction project.

LBP is more focused on overall productivity management rather than cost/time optimization only. It drives the improvement of the collective performance of the project rather than individual activities. Traditional highway construction planning uses LBP in the GIS-based platform. A more systematic and comprehensive use of LBP in GIS has the potential to increase understanding, accuracy, efficiency, and reliability of the construction plans and schedules. The location-based cost histogram is developed within GIS from location-based costs computed in the LBP process. It displays the breakdown of task costs at the location level. Location-level costs help in making location-based payments which ensure completeness of all tasks in all locations. The cost of each task in each location is computed and presented on a graph, defined as a location-based cost histogram. The availability of location-based costs facilitates location-based payment to contractors which ensures completion of all tasks in all locations (Sharma et al. 2014, 2016).

5.5.4.1 Location-Based Scheduling

LBP divides a project into locations instead of activities. LBP offers a new and comprehensive approach to planning and scheduling of infrastructure construction projects like highways, railways, canals, etc. (Kenley and Seppanen 2010). LBS represents the schedule in a time versus location graph. The schedule of a linear project in the time-location graph is usually presented with location on the x-axis and time on the y-axis. Tasks are plotted on the graph as lines running through various locations of a project. Each task is represented on the graph as a line with its inverse slope indicating the task progress rate – the rate at which the crews progress along the length in terms of meters per day. The slope may vary at various

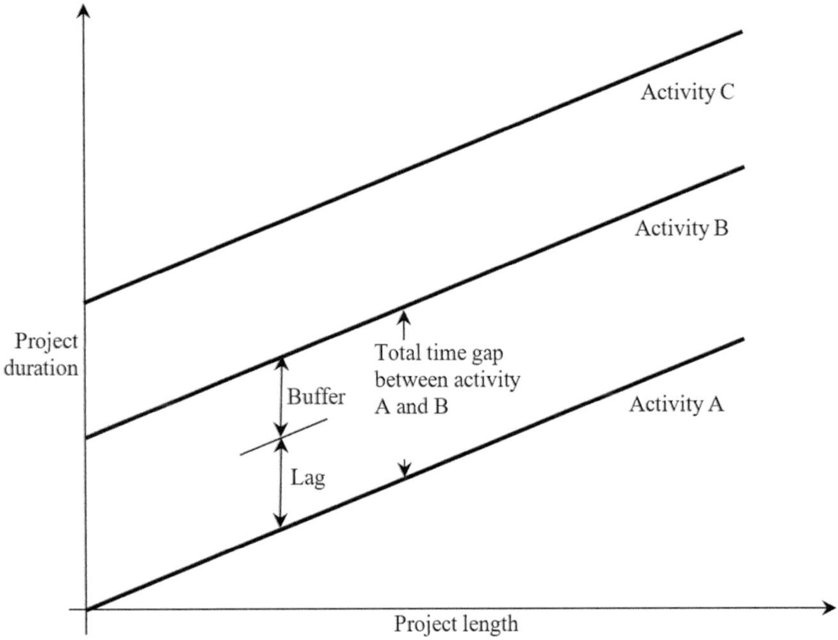

Figure 5.4 Schedule of a linear construction project in the form of a time-location graph

points, indicating varying progress rates due to variations in work quantities or task productivity. A balanced progress rate of all tasks, indicated by parallel lines, is ensured in LBS for efficient resource utilization, as shown in Figure 5.4. To achieve uniform production rates for all tasks, crew composition and number of crews are suitably adjusted (Sharma and Bansal 2018).

Synchronization and pacing are the two techniques that are utilized for ensuring smooth workflow and continuity of resources. Synchronization achieves a similar production rate for all activities, indicated by parallel lines. Pacing assigns the production rate to all tasks which ensures a continuous flow of crew from one location to another. This is achieved by selecting optimum crew sizes for various tasks. GIS helps in generating a time-location graph from information stored in tables within GIS. This representation provides quick visual information on workflows, productivity, and work locations which are critical in the execution of long-spanning projects like roads, canals, railways, laying pipelines, etc. LBS provides lead times, synchrony, concurrency, work-in-progress, sequence, direction, interfering trades, and buffers.

Accurate information about working locations is vital for efficient resource planning, safety on a construction site, and monitoring the progress of earthworks. LBP in GIS optimizes resource allocation when and where necessary and generates a location-based earthwork schedule. It also assists in the sequencing of cut/fill resources planning, identifying time-space congestion, monitoring activities progress from location aspects, and reducing gaps while communicating scheduling information among project teams. The location-based schedule in GIS in graphical format is easy to read and provides quick visual information about task production rates, work locations, work directions, and gaps between successive tasks. Easy visualization of such information helps in monitoring progress rates and making quick alterations for managing work or resolving conflicts. Hence, a linear graphical schedule in GIS is the best way to represent the schedule of repetitive linear construction projects. The availability of construction project information and execution schedule in GIS facilitates quick updating of the schedule whenever needed. Any change in spatial or non-spatial project information automatically updates the developed schedule. This saves the time and effort of planners and also ensures regular updates of the schedule (Sharma et al. 2014, 2016).

5.5.5 4D Modeling

In most of infrastructure construction projects, bar charts, CPM, and LOB technologies are widely used for planning and scheduling. Different activities in the schedule are related to one or more physical components of a project under consideration. The execution schedule provides non-graphical information which lacks in the graphical aspects of construction activities. To have graphical aspects of a project, a planner uses 2D drawings and associates the components of drawings with the related activities present in the schedule. A construction project may contain hundreds of activities; thus, interpreting a schedule without any link of its activities with the components of a project is a difficult task. This makes it difficult to check it for its completeness. The interpretation of the schedule also varies with the experience of the persons involved due to the difficulty in mentally linking each element in 2D drawings with the corresponding activities in the schedule (Koo and Fischer 2000).

These limitations of the schedule forced researchers to link it with 3D CAD components leading to the development of 4D models. A 4D model allows a planner to visualize the construction process in a way as it would actually be built. A 4D model increases the comprehensibility of the project schedule and allows users to detect potential problems such as scheduling conflicts before the construction. It assists a planner in avoiding space conflicts, examining constraints, and evaluating alternative construction methods. The potential of GIS for 4D modeling has been explored. GIS facilitates scheduling, sometimes the development of 3D components corresponding to different activities, importing 3D components from other widely used CAD or BIM-based tools, and linking activities of the schedule with 3D components. The components corresponding to each activity of a project are linked with the schedule using either on the earliest start time or the latest start time basis (Bansal and Pal 2008).

The components corresponding to different activities of the schedule are maintained in different layers in GIS. It is not necessary to have 3D components corresponding to each activity in the schedule. For example, activities like clearing and leveling of a construction site, marking a site, or curing of concrete does not have related 3D components. However, corresponding to each 3D component, there is an activity in the schedule. The number of layers developed corresponding to each activity depends on the degree of detail to be provided in a 4D model. The number of layers to be created corresponding to each activity may also depend on the shape, openings, and thickness at different levels of height. For example, as shown in Figure 5.5, one layer is enough to represent concreting in the foundation. However, the thickness of the foundation changes three times across the height, therefore, three different layers representing these changes in height may be created. Similarly, different layers of brickwork in the superstructure may be created at different levels of height to consider openings in walls. All components are linked with the schedule only if they comply with the requirement otherwise these are reshaped.

Sometimes, different layers may be merged together depending on the activities defined in the schedule. For example, in Figure 5.5, brickwork is shown with four different layers, if the entire brickwork is to be included in a single activity,

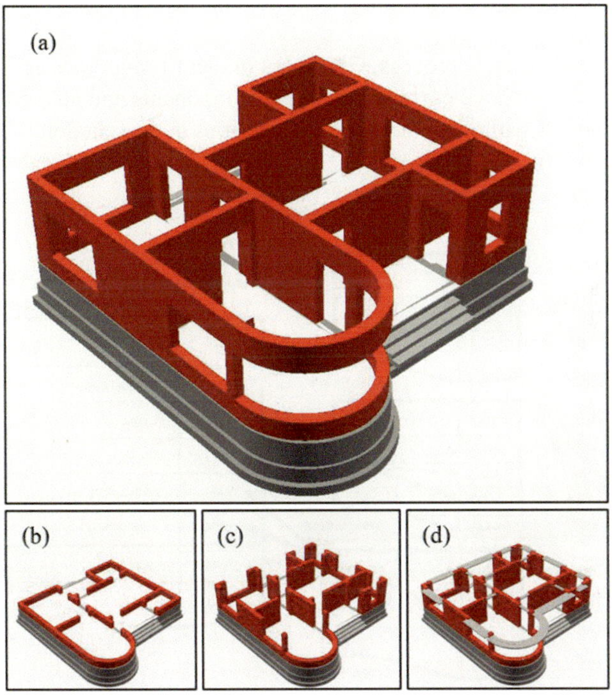

Figure 5.5 (a–d) The number of layers corresponding to an activity depends upon the complexity or degree of details required in a 4D model

four layers may be merged into a single layer. The components in a layer that belong to the same activity but are located at different positions in a 3D space may also be grouped together. For example, in Figure 5.5, the lintels over the doors and windows are not adjacent to each other but can be grouped into a single component.

The activities in the schedule are linked with the corresponding 3D components. The linkage between a component and an activity of the schedule may not always be one-to-one (one component corresponding to an activity of the schedule) always. However, relationships like many-to-one or one-to-many may also exist. In many-to-one linking, many components correspond to an activity in the schedule, as shown in Figure 5.6. In such a relationship, components may be merged together and linked with the corresponding activity in the schedule thereafter. In one-to-many linking, there is a single component corresponding to many activities. Linking involves adding a field *key* to the schedule table and the attribute table of each component. The field *key* is common between the two tables (i.e. schedule table and the feature attribute table of each component). It is used to establish connections between the components and the corresponding activities. All entries in the field key are done manually and are unique in both tables. Thus, the attribute required to associate the components with the corresponding activities in the schedule are the entries in the field *key*.

In the schedule review process, GIS helps in identifying those activities in the schedule that do not have corresponding 3D components and also provides a list of 3D components without corresponding activities in the schedule. The resulting

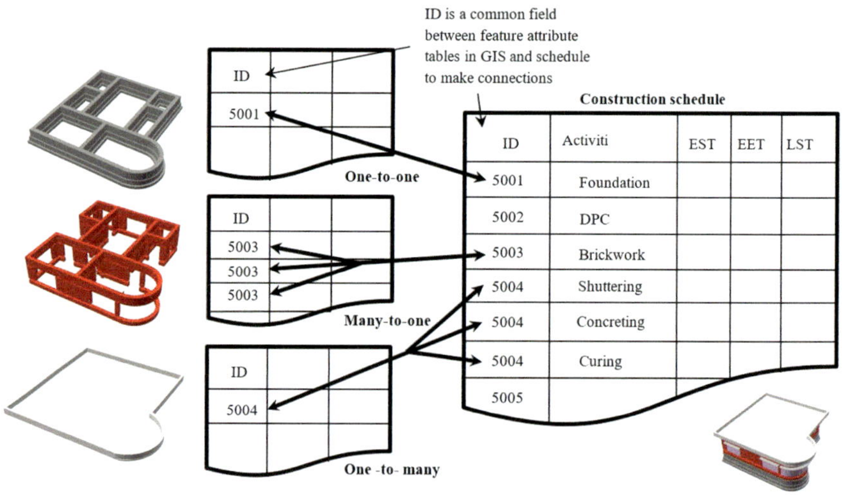

Figure 5.6 Different types of relationships used to link 3D components with the activities in the schedule

4D model is viewed for evaluating the schedule to verify the execution sequence, as shown in Figure 5.7. A 4D model displays at what time which components are to be built and where components will be in the space. This helps in understanding the 3D model and topological relationship between different components in many ways, like zooming, pan, fly forward, fly backward, etc. After analyzing the model, if it does not comply with the required execution sequence and needs some changes in the logic (predecessor relationships among various activities), the schedule is changed again as per the requirements. All activities which are not in a sequence are re-adjusted in the schedule. All relationships among the activities in the schedule that constitute a physical impossibility cause major disruption or delay at the construction site are corrected at this stage. After the implementation of the desired changes, the schedule is again associated with the related components for evaluation and sequence verification. That is, the 4D model is again developed by linking the components with the revised/updated schedule. It is finally accepted if no change in the logic and number of activities is required. If the numbers of activities in the schedule need addition/deletion or

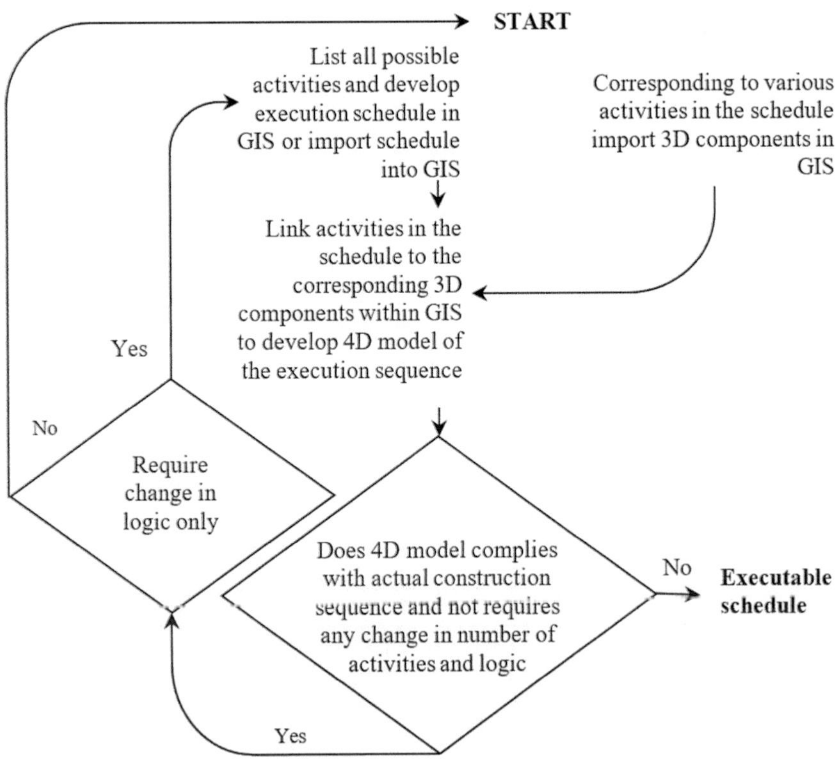

Figure 5.7 Procedure for visualization and evaluation of the construction schedule in GIS

if some editing is required in the corresponding components, the entire procedure is repeated again (Bansal and Pal 2011).

5.5.5.1 *Integration of 4D Model with Its Surroundings*

Chapter 3 covers the modeling of surroundings. 4D model of the execution sequence is enriched by integrating it with its surroundings (nearby existing facilities/utilities, water supply lines, road network, power supply lines, land-use, etc.). The 4D model of the execution sequence along with its surroundings is shown in Figure 5.8. The virtual model of another building along with its surroundings which is planned to be constructed, is shown in Figure 5.9. This allows the realistic visualization of the execution process along with its immediate surroundings. GIS provides a data interoperability extension to directly read/write or translate different spatial data formats to import 3D components in GIS or export 4D models from GIS.

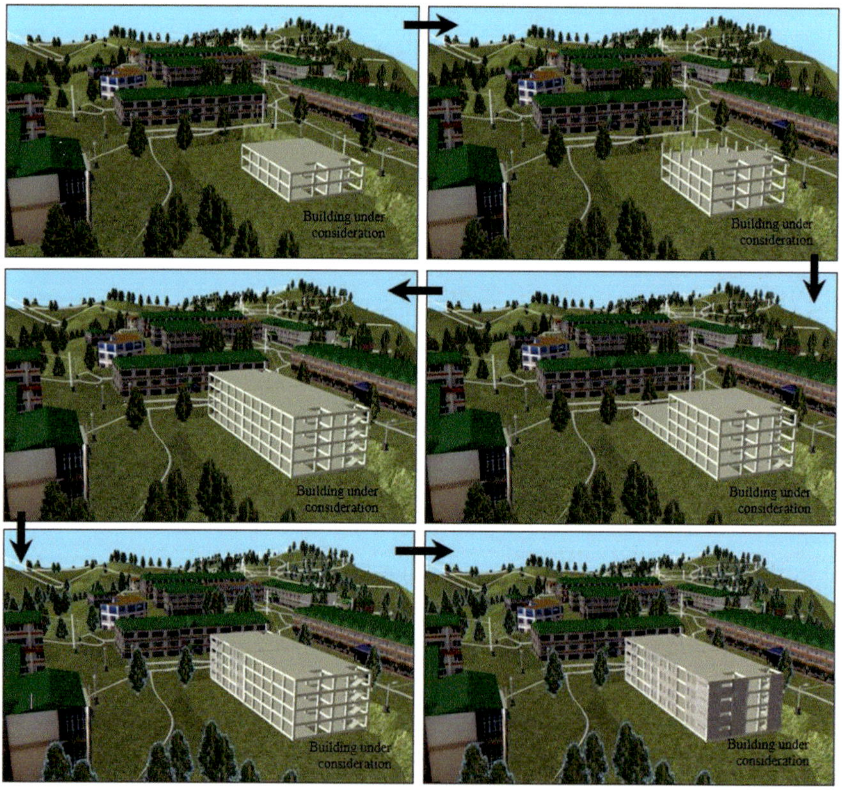

Figure 5.8 4D model of the execution sequence along with its surroundings

Figure 5.9 The virtual model of a building along with its surroundings

5.5.5.2 Degree of Details in 4D Model

In order to have sequential visualization in the 4D model, the schedule should have a sufficient degree of detail in its work breakdown structure. In case some activities in the schedule are with a limited degree of detail, the developed 4D model does not communicate the execution process in a desired sequential manner, even if the schedule is correct. For example, in Figure 5.5(a), in the load-bearing construction, activity brickwork in the superstructure appears after the construction of the foundation. At that time no door or window frames were fixed on the place. Thus, the brickwork shown in Figure 5.5(a) should not appear complete unless all door and window frames are fixed. Similarly, lintels over the doors and windows should also appear before the complete appearance of the brickwork. This shows the physical impossibility reflected by the 4D model. Therefore, the current level of detail for this activity is not sufficient to describe the execution sequence correctly. The brickwork that appears in the 4D model, as represented in Figure 5.5(a), lacks in the degree of detail. The schedule should have sufficient numbers of activities and sub-activities to properly represent the hierarchy of work breakdown structure. Some activities of the schedule may encompass a set of sub-activities that represent each activity in more detail, thus, helping to understand the logic or execution sequence correct way. For example, to address such issues in detail, the activity brickwork in the superstructure is divided into sub-activities at different levels of height. Figures 5.5(b, c, and d) represent the sequence of the construction in the desired manner.

5.5.5.3 Advantages of 4D Modeling in GIS

GIS utilizes 3D components generated in CAD or BIM-based software tools to develop the 4D model. GIS is used to develop an execution schedule and also reads the schedule generated in commercially available scheduling tools. In case, the resulting 4D model does not comply with the execution sequence, the schedule generated in GIS is easily corrected in GIS itself but the schedule generated in commercially available scheduling tools cannot be corrected in GIS. Further, the availability of a large number of activities in a project network makes it difficult to confirm that all components of the project have related activities. To ensure this, GIS helps in preparing the list of activities in the schedule without corresponding components. GIS also lists down all components of a 3D model that do not have corresponding activities. In addition to this, a visual check of the schedule is also possible by viewing it graphically in 3D, which may help in preventing the omission of any activity in the schedule.

The visualization of the execution sequence displays which component is to be built at what time and where in a geographical space. The components that are scheduled or planned to be in operation on and before a date are graphically displayed in a virtual geographical space. This allows checking of the schedule for each calendar date. The users have the option to rotate 3D components around the x, y, or z-axis to observe the developed 3D models from any direction and angle. The failure or success of an infrastructure project depends on the quality and timely availability of the information from the database. GIS allows for maintaining spatial and non-spatial information together in a single database. The non-spatial data includes construction resource data, specifications for quality control, safety recommendations, etc. When a component corresponding to an activity is clicked, GIS informs users about the various times, available floats, and resource information. GIS-based animations of the execution sequence allow detection of accident-prone locations and deciding preventive measures (such as placing warning signs, restricting access, providing safety guards, etc.). By viewing the time and location of the work, respective workers are informed about the possible hazards.

Detection of logical errors and schedule correction: A Logical error is the missing of relationships between two or more activities or an incorrect execution sequence among various activities. This constitutes a physical impossibility that may cause major disruption and delay at the construction site when actual execution starts. Logical error is removed before finalizing the schedule. The doubtful execution sequence is checked on a computer screen with the help of 4D modeling in GIS. Before finalizing the execution schedule, missing activities are inserted and activities that are not in sequence are rearranged suitably. If some discrepancies exist in the components, which together constitute a complete 3D model of an infrastructure, the 3D components are corrected.

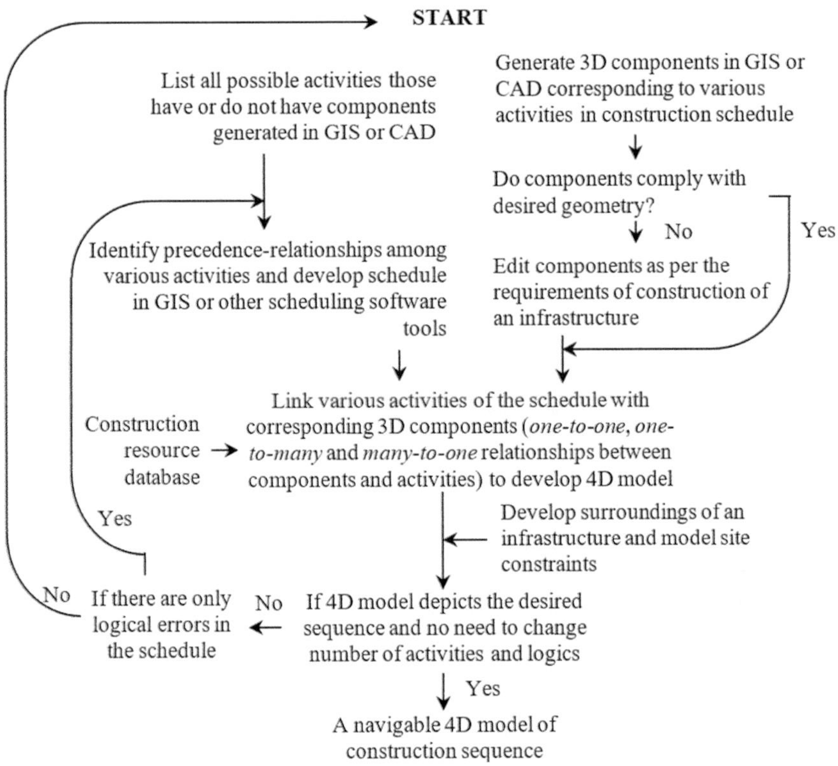

Figure 5.10 Animation-based construction schedule review and correction process

After implementing the desired changes, the schedule is again linked with the components so as to generate the correct execution sequence in the form of animations.

Missing activities in the schedule: 4D modeling is used to detect missing activities in the schedule. It is possible to find a 3D component corresponding to that there is no activity in the schedule. It is also possible to find an activity in the schedule corresponding to that there is no 3D component. Figure 5.10 shows the animation-based construction schedule review and correction process used in GIS.

5.5.6 *Resolving Workspace Conflicts*

Construction involves many subcontractors who work within a limited space. Each subcontractor needs a job-specific workspace, equipment space, material storage space, travel paths, etc. Deficiencies in space planning result in a congested site, loss of productivity, space conflicts, schedule interference, and project delays. Such problems result in an uncontrolled site and disputes among subcontractors sharing a common site space. It is impractical for a planner to mentally plan for multiple types of space requirements which it change with time/schedule like any other resource requirements. The virtual visualization of the execution sequence is widely used in space planning. The construction space planning is not only related to the sequence visualization developed in 4D modeling tools. For example, space planning for gravity dam construction, where topography plays a major role cannot be simulated without geospatial capabilities which are missing in 4D tools (Bansal 2007).

Other factors like environmental condition, site topography, thermal comfort, access route planning, etc., which are not modeled in CAD-based 4D tools, however, influence space requirements. CAD-based 4D tools lack in geospatial analyses like: evaluation of a site with respect to flooding, site drainage planning in case of flooding, route planning of vehicles carrying consignments from different access routes to a site, etc. GIS is a powerful technology to make decisions in cases where a site is large and complex in its topography. 3D virtual model, surrounding topography, 4D modeling, and geospatial analysis capabilities together in a single GIS platform help in space planning much better way.

5.5.6.1 4D Animation of Workspace Requirements

Generally, three categories of spaces are considered in space planning which are: site space on the ground, space provided by temporary structures such as scaffolding or working platforms, and space provided by a structure constructed over time. The categories of available spaces are described in terms of sizes, locations, and times of availability. A single activity consists of several jobs that require working and path spaces for labor, equipment, and materials storage; hence, various categories of space requirements for each activity are listed. The spaces required may be located outside, inside, above, below, or around the reference component to be constructed. The shape of a space required may be cubical, spherical, or prism. GIS is used to virtually represent the activity-specific space requirements after inputting the space descriptions. However, the activity-specific space requirements on a site change with the execution schedule. GIS is used to link the space requirements with the execution schedule to generate dynamic space requirements. Space-loaded 4D animation is generated in GIS to represent the space requirements of various activities along with 3D components to be constructed, as shown in Figure 5.11. In space planning, to finalize an execution plan in terms of when, where, and how long a space is required on a site, space-loaded 4D animation is quite helpful. The overlaps among various spaces are verified visually through space-loaded 4D animations (Bansal 2011a).

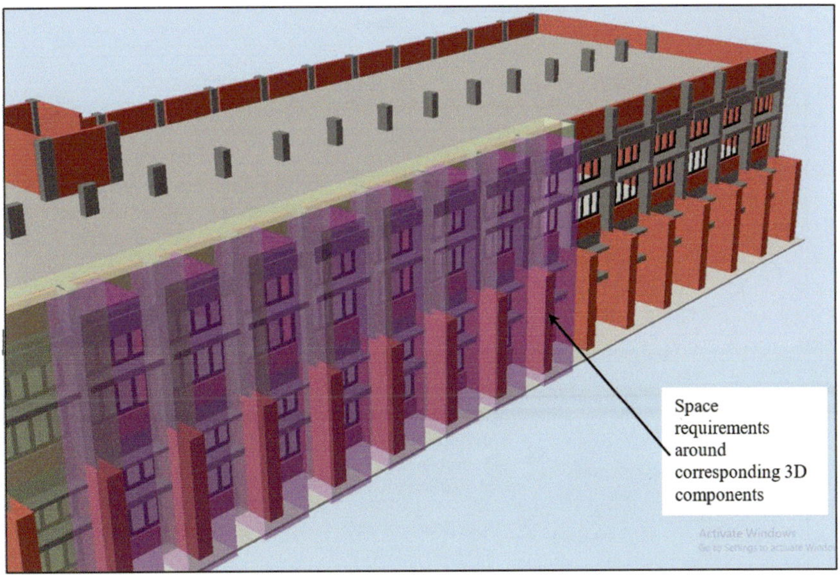

Figure 5.11 A screenshot of space-loaded 4D animation used to represent the space requirements of various activities along with corresponding 3D components

5.5.6.2 Workspace Conflicts Identification and Resolution

The detailed procedure for the identification and resolution of space conflicts used in GIS is shown in Figure 5.12. GIS uses 3D animation and topology to check time-space conflicts. A space conflict is said to occur when a space during the same time period is demanded by more than one activity or/and when spatial relations are violated. Once a space conflict is detected, the affected activities are identified as well as the sizes of conflicting spaces are estimated. The following conflict resolution strategies are used to either resolve or minimize space conflicts:

- Initially, to resolve space conflicts, the start or finish times of activities participating in conflicts are changed depending upon the availability of total floats. The effects of changes in start and finish times on the construction sequence and space requirements are previewed with the help of space-loaded 4D animation in GIS.
- In case the use of float does not help in resolving space conflicts, a conflict resolution strategy is selected. The space demands are adjusted by changing the locations of conflicting spaces or by dividing the originally assigned spaces into small parts. In this strategy, the space requirements for material storage, operation yards, etc., are altered to eliminate or minimize space conflicts.
- Concurrent activities in the schedule sometimes lead to site congestion, schedule delays, and an unsafe working environment. In such cases above strategy may not

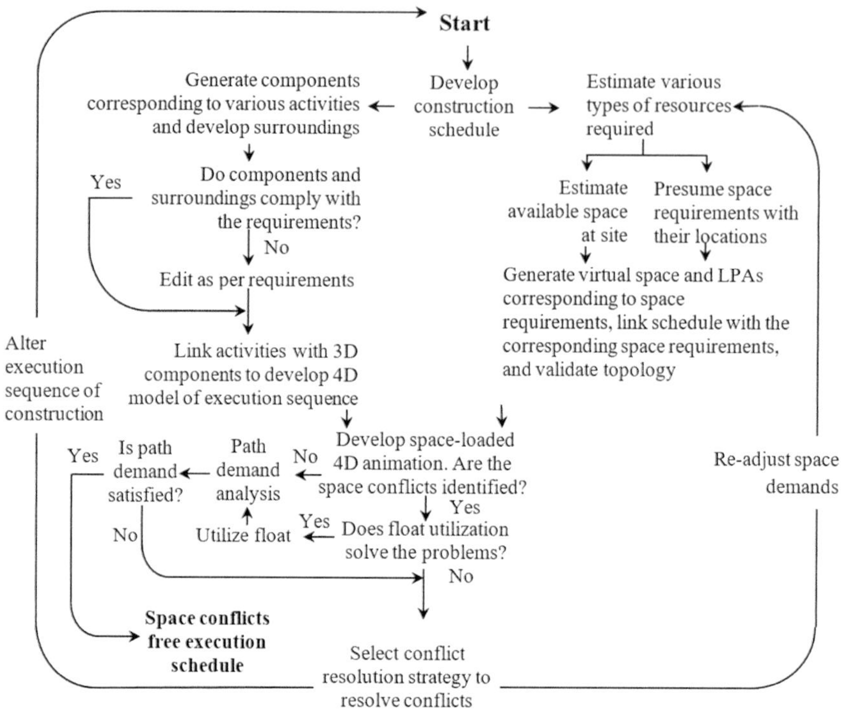

Figure 5.12 Construction space planning to develop a space conflict-free execution schedule

work. Therefore, such situations are avoided by correcting the execution schedule before its implementation. Alternate construction sequences are explored by re-sequencing project activities or by changing construction methods. This includes changes in precedence relationships among various activities, durations of space requirements, or splitting a few activities to resolve space conflicts.

- The space requirements, time duration of activities, and execution schedule are changed at the same time, as the last alternative. The changes in space requirements, construction methods, or/and sequence most likely resolve the majority of space conflicts. In the sequence alteration process, activities in the schedule need addition/deletion and editing in the corresponding components' geometry. The revised schedule is again linked with the corresponding components after implementation of desired changes and the space requirement is again reviewed (Bansal 2011a).

5.5.7 Planning for Construction Safety

Workers in the AEC industry face a greater risk of fatality or injury than workers in any other industry; therefore, their protection is of great concern. Effective safety planning contributes towards the prevention of accidents and ill health of site personals. The safety planning in the AEC industry is generally done separately from project execution planning; however, there must be a link between the two. There are two reasons behind the importance of this link. First, safety engineers need to identify *when* and *where* safety measures are required. Secondly, because design drawings/procedures have the information related to *why* and *what* safety measures are needed. Project execution planning and safety planning together, convey *what* is to be built and *what* safety measures are necessary *when*, *where*, and *why*.

Safety planning involves the identification of all potential hazards and accordingly deciding on safety measures. Safety planners visualize 2D drawings into a 3D model and mentally link components with the respective activities in the schedule to understand the execution sequence for safety planning. Sequence interpretation and accordingly hazard identification may vary with the level of experience, knowledge, and individual perspective of a safety planner. Therefore, 4D modeling is used to create the simulation of the execution process by linking the execution schedule with a 3D model of an infrastructure. However, in 4D modeling, an infrastructure project is modeled in isolation, that is, without considering its immediate surroundings which affects the safety planning. 4D modeling, modeling of surroundings, and geospatial analyses together in GIS help in the development of a safe execution sequence. A 3D model of an infrastructure along with its surroundings is linked with the execution schedule to develop a 4D model for safety planning. During the safety review process, if the planned sequence results in a hazardous situation, it may be corrected before its actual implementation (Bansal 2011b).

GIS is also used for the development of a safety database. The safety database in GIS is developed by soliciting safety recommendations from safety professionals, safety procedures, and related safety codes. The database is easily processed to determine what types of safety measures are possible. The database is kept safe and made available to those who updates it regularly.

5.5.7.1 Linking Safety Database with Construction Activities

The safety information is retrieved from the safety database and linked with the various activities of the schedule. This facilitates in display of the requested safety information when needed by avoiding too much information being explicitly displayed. It is also possible to link specific procedures of a company with the corresponding activities to handle project-specific operations. The primary role of linking procedures and safety measures with corresponding activity is to ensure compliance with safety regulations and get help in deciding safe construction methods.

5.5.7.2 Linking Safety Database with 3D Components

The components of a 3D infrastructure model are also linked with the safety database. In this process, three types (*one-to-one*, *many-to-one,* and *one-to-many*) of links between components and linkage files are used. In a *one-to-one* link, a component of an infrastructure model is connected with a single linkage file. The linked linkage file connects component with corresponding safety documentation files in the safety database maintained in GIS. In the case of a *many-to-one* link, several components correspond to a single linkage file. The linkage file connects components with the safety documentation files of the safety database. In the case of a *one-to-many* link, a single component corresponds to the number of linkage files. The linkage files connect a component with many safety documentation files in the safety database. A component of a 3D model corresponds to one or more files in the safety database, depending upon the type of link (Bansal 2011b). GIS provides a user-friendly method to retrieve safety measures for each component of a 3D infrastructure model. Figure 5.13 illustrates the construction safety planning process for reviewing of execution schedule for developing a safe execution sequence.

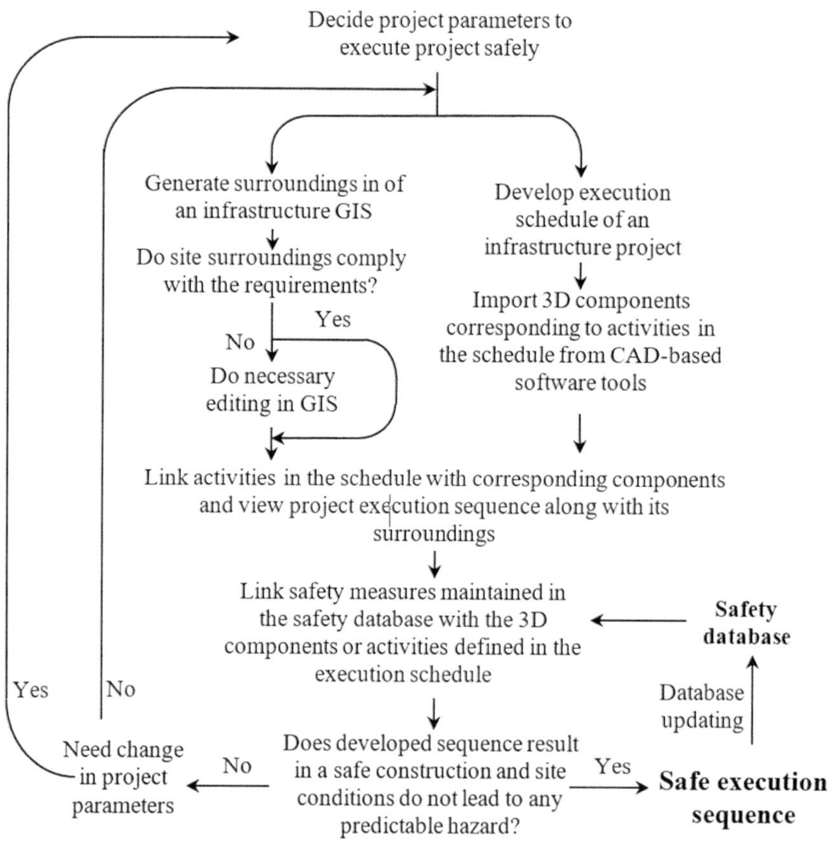

Figure 5.13 Construction safety planning process: review of execution schedule for developing a safe execution sequence

A 4D model of the execution sequence in GIS helps in the identification of hazardous situations and determines *when* and *where* safety measures are needed. GIS generates and maintains spatial information regarding geographical locations and site topographical conditions to answer *why* in safety planning. In the safety review process, if the planned sequence does not appear safe, an alternative construction sequence is explored by re-sequencing project activities, addition/deletion of activities, editing components' geometry, or by changing construction methods. After the implementation of the desired changes, a schedule is again linked with the corresponding components for the safety review. It is an iterative process until safety professionals are fully satisfied. The sequence is finally accepted if safety professionals approve it.

GIS facilitates the analysis of execution sequence, site analysis, daylighting, and predicting accidental situations to demonstrate how accidents can be prevented from happening. Using the GIS environment, safety planners and other project participants match past accidents which were already occurred under conditions that are similar to the ongoing project and prevent the occurrence of similar accidents again in the future. GIS-based safety planning process facilitates an easier understanding of the construction sequence and predicts places and activities that have a higher potential for accidents. GIS-based safety planning along with geospatial analysis capabilities, has its own strength where the lifecycle of an infrastructure is perceived together with its surroundings. GIS is also used for safety monitoring in the construction planning of hydropower projects (Jia et al. 2012). GIS is used for the development of a web-based spatial decision-support system for proactive health and safety management for linear construction projects (Atay and Toz 2011). Linear projects are scattered over large distances; GIS helps in identifying potential hazards and their locations using topographical and layout maps of the construction site and offers mitigation measures.

5.5.8 Developing Complex Visualization

Construction planning does not only depend upon the visualization developed in through 4D CAD, VR, construction simulation, or even planning information integration in BIM. For example, planning of a dam construction where topography plays a major role cannot be simulated without geospatial analysis capabilities which are missing in 4D CAD, VR, or BIM (Dollner and Hinrichs 2000). GIS helps in the development of 3D visualization of complex infrastructure construction planning (Bansal and Pal 2007; Zlatanova and Prosperi 2005). The complex 3D visualization is also used for security and emergency planning. GIS allows importing information from numerous software tools which in turn allows visualization and spatial query in two or three dimensions. The construction process simulation is an effective tool for infrastructure construction planning and improving performance during the construction stage. The simulation tools lack the capabilities to represent explicit geographic information in the simulated construction planning process. GIS is utilized to overcome this limitation by depicting simulated operations dynamically in a 3D environment (Zhong et al. 2004). GIS offers powerful planning, visualizing, and querying capabilities as well as facilitates in the detection of logical errors in the simulation models. It helps in the selection of an optimal construction method after simulating the potential construction methods in the construction planning stage. Visualization in GIS is helpful in detecting performance inconsistencies and obtaining insight into the simulated construction operations. GIS helps in developing geographical visualization for planning and monitoring of construction projects that are repetitive in nature (Sonmez and Uysal 2008).

5.6 BIM in the Construction Planning Stage

BIM has brought a transformative shift in the way projects are planned and managed within the AEC industry. Moving beyond traditional methods, BIM introduces a data-rich, integrated approach that enhances both the accuracy and efficiency of project planning. One of the most significant advantages of BIM lies in its visualization capabilities and parametric modeling features, which allow stakeholders to interact with a virtual representation of the project in real-time. These features improve collaboration, reduce errors, and support more precise decision-making in the infrastructure project planning stage. BIM has proven particularly valuable in construction sequencing and scheduling, allowing for the generation of 4D models that integrate the construction schedule with building components. This enables project managers to visualize the construction process over time, identify potential scheduling conflicts, and make adjustments before work begins on-site. BIM is also instrumental in workspace planning, ensuring that different activities in the schedule operate without spatial conflicts and logistical interference. Additionally, BIM plays a critical role in work execution planning and site layout design, offering detailed simulations that optimize the arrangement of materials, equipment, and workforce on a construction site (OpenAI 2025).

Beyond the immediate scope of construction planning, BIM also supports broader logistical and operational tasks. For example, it can be utilized for logistics planning and supply chain optimization, ensuring that materials and components are delivered efficiently and stored safely, thereby minimizing delays and reducing costs. Furthermore, BIM contributes to sustainability efforts by reducing material wastage, improving resource allocation, and supporting lean construction practices through accurate quantity takeoffs and real-time data coordination. The subsequent sections of this chapter delve deeper into the diverse applications of BIM during the construction planning stage, with particular attention to how it facilitates more informed, efficient, and environmentally responsible infrastructure development. Through these applications, BIM not only enhances the construction planning part but also contributes to a more sustainable and integrated approach to project delivery in the AEC industry:

5.6.1 *BIM for Scheduling Construction Projects*

The development of an execution schedule is critical in the successful delivery of projects within the AEC industry. It serves as a tool for managing time, cost, and quality throughout the construction process. However, the development of an accurate and executable construction schedule remains a persistent challenge. Traditional scheduling approaches rely heavily on the expertise and intuition of planners, particularly their knowledge of work packages, prior experience with similar projects, and their capability to foresee interdependencies and site conditions. Moreover, creating a viable construction sequence demands a deep understanding of project geometries and the internal relationships among building components. BIM addresses many of these challenges by offering a rich source of project data that can be utilized to support the generation of construction sequences. BIM models contain detailed geometric and structural information, which when appropriately extracted, can significantly enhance the precision and reliability of developing execution schedules. Despite its potential, formal control mechanisms for ensuring the completeness and correctness of execution schedules remain limited in traditional methods. Scheduling is often performed manually, which is time-consuming, inconsistent, and prone to error. In contrast, BIM-integrated scheduling, particularly when coupled with 4D visualization, allows for the automated generation of construction sequences. Because BIM objects are logically connected to construction tasks, dynamic and flexible schedule is generated with minimal human intervention, offering visual representations of the construction process that aid in both planning and communication (Tauscher et al. 2014).

BIM is used to automatically generate on-site construction schedules by considering structural support systems and the topological relationships among building components. Through algorithmic processing, precedence relationships between building components are derived and used to create a logically ordered construction sequence (Liu et al. 2014). This capability is further enhanced by integrating BIM models with different activities in the schedule, process simulations, and advanced optimization algorithms, allowing for the generation of activity-level construction schedules that are not only automated but also optimized under resource constraints (Liu et al. 2015). Such integration overcomes many of the shortcomings of conventional scheduling practices by offering a more holistic, data-driven approach to time management.

Traditional scheduling methods, such as the CPM and PERT, often fall short in terms of precision due to their inability to accurately reflect the dynamic and uncertain nature of construction environments. To mitigate these shortcomings, researchers and practitioners are increasingly combining BIM with discrete event simulation and advanced planning techniques such as takt time. Discrete event simulation enables the modeling of construction as a dynamic system, allowing for the simulation of resource flows and time-dependent processes. When integrated with BIM and takt time planning, it becomes possible to implement just-in-time strategies, resulting in realistic and optimal construction schedules that reduce waste and enhance efficiency (Abbasi et al. 2020).

The genetic algorithms are used to automatically derive structurally stable construction sequences by evaluating multiple construction order scenarios and selecting the most feasible solution based on predefined criteria (Faghihi et al. 2014). BIM further supports site-level operation simulations, providing insights into material handling, equipment movement, and managing workforce logistics, which are essential for generating detailed and actionable project schedules (Wang et al. 2014). Beyond scheduling, BIM is also being leveraged for risk management, especially in large and complex infrastructure projects. Its integration into established construction management workflows enhances decision-making by aligning model functionality with key processes such as procurement planning, hazard mitigation, and contingency management (Hartmann et al. 2012). BIM has revolutionized the construction scheduling process by automating sequencing, enabling resource-aware planning, and supporting simulation-driven optimization. When combined with advanced computational methods such as discrete event simulation, takt time, and genetic algorithms, BIM facilitates the creation of execution schedules that are realistic, adaptive, and aligned with project goals, thus addressing many of the limitations inherent in traditional scheduling practices.

5.6.2 Resolving Workspace Conflicts

In the infrastructure construction projects, available workspace for work execution is recognized not only as a critical resource but also as a constraint that must be effectively planned during the project scheduling phase. Inadequate workspace planning often leads to time-space conflicts, where overlapping tasks or limited physical workspace hinder workflow efficiency. The workspace conflicts disrupt construction sequences and cause significant time delays. To address this issue, BIM is employed to detect and resolve workspace conflicts. 4D-BIM that integrates the 3D model with schedule is particularly effective in dynamically identifying and quantifying time-space clashes and their potential impacts on project performance (Hasannejad et al. 2023).

Workspace conflicts arise when multiple construction activities compete for the same workspace within the same timeframe, affecting crew movement and overall efficiency. BIM tools facilitate the detection of such overlaps by visualizing the temporal and spatial occupation of each activity, thereby enabling planners to anticipate and mitigate disruptions. The severity of conflicts is quantitatively assessed based on their impact on productivity, allowing for more precise conflict resolution and planning adjustments (Mirzaei et al. 2018).

A major contributing factor to workspace clashes is the siloed planning of different construction disciplines, like civil, mechanical, electrical, plumbing, etc., often conducted without integrated coordination (Akponeware and Adamu 2017). BIM overcomes this limitation by automatically detecting interferences across different disciplines in real-time. For example, it ensures that mechanical components like pipes and ducts do not intersect with structural elements such as steel, beams, or walls. Through 3D design coordination, BIM significantly reduces physical clashes during the designing and construction planning stages, minimizing rework during construction and leading to cost and time savings. BIM enhances the clash detection process, enabling stakeholders to perform cost-benefit analyses, prioritize high-impact conflicts, and integrate lessons learned into future projects (Chahrour et al. 2021).

Time-space conflicts also have direct implications for on-site safety and productivity, particularly in highly congested work environments. To tackle these challenges, researchers have proposed hybrid conflict management approaches that integrate BIM with discrete event simulation and rapidly-exploring random tree path planning algorithms. These advanced systems automatically detect, assess, and resolve all types of time-space conflicts throughout the project lifecycle, based on predefined logical rules. This automation enhances conflict mitigation efforts and supports safe and uninterrupted project execution (Dashti et al. 2021). Beyond physical clashes, disputes in the AEC industry are a major contributor to poor project performance. BIM also plays a vital role in identifying potential sources of dispute early in the project. Techniques such as the analytical hierarchy process and multi-attribute utility theory are employed to develop decision-making frameworks for managing causes of disputes (Charehzehi et al. 2017).

5.6.3 *Construction Site Layout Planning*

Site layout planning plays a critical role in construction projects and is typically carried out directly on-site to determine the most efficient arrangement of temporary facilities, materials storage, site offices, etc. The primary objective of site layout development is to minimize the travel distances for personnel, materials handling, and equipment positioning, thereby improving overall site productivity and safety. This process involves the development of dynamic layout models that reflect the dynamic nature and requirements of a construction site throughout the construction stage. However, developing a layout is often a labor-intensive and time-consuming task. It demands extensive manual input, frequent updates in response to design changes, and integration with constantly shifting construction schedules and plans. As a result, traditional methods are inefficient and prone to errors. To address these challenges, BIM is an effective approach to site layout planning. BIM enhances both the collection and processing of qualitative and quantitative data. BIM assists in obtaining accurate quantitative data, such as spatial dimensions, quantities, and scheduling, qualitative insights are derived using rule-based systems grounded in expert knowledge (Le et al. 2019). Furthermore, BIM supports model-based rule checking, which enables planners to validate and optimize site layout decisions systematically. When combined with advanced optimization techniques like genetic algorithms, BIM is used to create automated frameworks that generate dynamic site layout models. This integration significantly improves planning accuracy and adaptability (Kumar and Cheng 2015).

5.6.4 *Simulation of Equipment Use*

BIM is increasingly utilized to optimize the placement of heavy construction equipment, such as tower cranes, on complex or congested construction sites. The accurate positioning of such equipment is crucial for maximizing operational efficiency and minimizing construction delays. To achieve this, project planners formulate the equipment positioning as an optimization problem, often with the objective of reducing the total cycle time of operations, such as material handling or lifting tasks. Depending on the specific requirements and complexity of the construction site, different optimization techniques are applied to identify the most effective configuration (Dasović et al. 2019). Beyond equipment placement, BIM also plays a significant role in optimizing the positioning of surveillance and monitoring systems, such as construction cameras. These cameras are essential for enabling autonomous site monitoring, enhancing safety oversight, and ensuring progress tracking. However, determining optimal camera locations is a non-trivial task, particularly on construction sites that are dynamic and constantly evolving, with dynamic obstacles and changing work zones. BIM addresses this challenge by providing an accurate and up-to-date digital representation of the site, which allows planners to evaluate and adapt camera placement throughout various phases of construction. This capability makes it possible to maintain effective visual coverage despite continuous changes in a site environment (Chen et al. 2021). In the absence of BIM, identifying and maintaining optimal configurations for both equipment and monitoring systems becomes significantly more difficult, especially in dynamic and high-complexity construction settings.

5.6.5 Planning for Safety During Construction

Worldwide global occupational safety statistics consistently highlight that the construction industry records some of the highest accident rates across all sectors in many countries. Despite the existence of stringent legislation and safety regulations, violations remain widespread and concerning. The primary reasons behind this ongoing safety crisis include a lack of timely access to critical safety information and the absence of integration between safety data and project-related activities in the schedule. BIM has emerged as a powerful tool with the potential to significantly enhance construction safety performance. Among its many features, visualization capabilities stand out as especially promising, enabling a clearer understanding and communication of safety risks. One of the most impactful applications of BIM in this domain is hazard identification, which plays a central role in improving on-site safety conditions (Akram et al. 2019).

BIM is increasingly being used to develop automated systems capable of analyzing building models to detect potential safety hazards and recommend appropriate preventive measures. These systems proactively identify and address safety issues during both the design and construction planning phases. For example, automated safety-checking platforms within BIM inform engineers and site managers about the nature, location, timing, and necessary preventive actions for fall-related risks, one of the leading causes of construction fatalities, before physical work begins (Zhang et al. 2013). Moreover, BIM is extended to assess occupant safety through design validation. In this approach, predefined safety rules and standards are encoded as a rule-set within the BIM model. The system then automatically identifies any building components that violate these codes, allowing designers to correct them early in the design or construction planning stage. This rules-based checking method helps prevent safety hazards not just for workers, but also for future occupants of a building (Parsamehr et al. 2023).

BIM also supports activity-level site planning with a proactive focus on safety. Integration with remote sensing and workspace modeling technologies enables automated visualization of active work zones. For instance, by using global positioning system data loggers attached to workers' hard hats, construction teams can track movement and activity patterns in real-time. These data points are used to extract activity-specific workspace requirements, which are then visualized on the BIM platform. This allows project stakeholders, including engineers, planners, site managers, foremen, and workers, to identify potential workspace conflicts between different crews or between workers and heavy machinery. Such insights enable better decision-making regarding safe work sequencing and spatial coordination, reducing the risk of accidents while enhancing overall productivity (Zhang et al. 2015).

Given the variation in safety standards across countries, ensuring compliance is a complex challenge. To address this, customizable BIM-based rule-checking platforms have been developed. These platforms apply adaptable rule-based algorithms within BIM software tools to detect safety violations during the early stages of design and planning. Falls remain one of the most serious and frequent hazards on construction sites, often resulting in severe injury or death. When these hazards are identified through BIM, safety equipment is designed and integrated into the construction schedule in advance. BIM thus enhances the traditional safety planning process by facilitating more accurate and timely decisions regarding fall protection systems (Melzner et al. 2013).

Furthermore, safety analysis is conducted as part of a dynamic simulation of the entire construction process. By combining BIM with 4D modeling and construction simulation technologies, project teams evaluate safety conditions over time, taking into account the evolving structure and site layout. This integrated approach provides a more practical and comprehensive framework for time-dependent safety assessments, making safety planning a continuous and responsive process rather than a one-time task (Hu et al. 2008).

5.6.6 Logistics and Resources Planning

Effective logistics and resource planning are fundamental to achieving critical project objectives, particularly those related to cost efficiency, timely delivery, and on-site safety. In construction, where operations are complex and constantly evolving, logistics management ensures the smooth flow of materials, equipment, and workforce, reducing delays and minimizing waste. The integration of BIM with lean construction principles provides a powerful framework for managing these complexities. Lean philosophy emphasizes the elimination of non-value-adding activities, continuous improvement, and efficient resource utilization. When combined with BIM's data-rich and visual modeling capabilities, this approach offers enhanced visibility and control over logistics processes throughout the construction stage. The application of 4D BIM allows for detailed simulation and planning of construction logistics. This enables project teams to visualize the delivery and movement of resources in sync with the construction schedule. Logistics planning is performed at multiple organizational levels, from strategic and tactical to operational, allowing for better alignment of logistics decisions with project milestones. Moreover, this integrated approach is effectively synchronized with the last planner system, a lean production control methodology focused on improving workflow reliability and collaborative planning. By aligning logistics planning with the last planner system, 4D BIM supports short-term work planning, enhances communication among stakeholders, and reduces uncertainties in material delivery and resource availability (Bortolini et al. 2019). This synergy results in a more predictable, efficient, and responsive construction process.

5.6.7 *Supply Chain Optimization*

Prefabrication and modular construction techniques play a significant role in reducing both the overall costs and the duration of on-site activities in construction projects. By manufacturing components off-site in controlled environments, these techniques help streamline the construction process and minimize disruptions typically associated with on-site work. When integrated with BIM, these techniques become even more effective. BIM allows for the synchronization of the on-site construction schedule with the manufacturing or fabrication timelines of various supply chain participants. This alignment leads to a significant reduction in total construction time by optimizing the coordination between site operations and off-site production. In particular, careful planning of on-site activities, combined with detailed knowledge of lead times required for prefabricated elements and the logistics of transporting them, contributes to cost savings across the entire supply chain. BIM facilitates the integration of critical flows, including information, materials, and cash, as well as transportation logistics, to create an optimized just-in-time delivery schedule. This optimized delivery plan ensures that prefabricated components arrive at the construction site precisely when needed, avoiding both idle time and storage issues (Ocheoha and Moselhi 2018). Just-in-time scheduling approach is especially valuable for large and complex projects, where discrepancies between off-site and on-site timelines can significantly affect productivity. BIM enables project teams to anticipate such variations and forecast accurate delivery dates for precast elements, thereby maintaining workflow continuity and reducing the risk of delays.

BIM platforms are increasingly being used to manage and enhance various operations across the construction supply chain. One notable application is within the reinforcement supply chain, particularly for concrete reinforcement. BIM has proven highly effective in establishing seamless information flow processes throughout this supply chain (Aram et al. 2013). BIM supports the planning, fabrication, delivery, and installation of reinforcement components, ensuring improved coordination, transparency, and overall efficiency. By enabling real-time data exchange among stakeholders, BIM helps optimize procurement, reduce waste, and enhance the overall performance of the reinforcement workflow. The integration of BIM with prefabrication and modular construction methods not only shortens project timelines and reduces costs but also enhances collaboration and operational efficiency across all stages of the construction supply chain.

5.6.8 *Planning for Reduction in Construction Waste*

BIM plays a pivotal role in planning for reducing construction waste, particularly during the early construction planning stage. At this stage, several factors are identified that to contribute significantly to construction waste, including design and documentation errors, poor material specifications, inaccurate cost estimations, site condition misjudgments, inadequate work planning, and procurement inefficiencies. BIM serves as a powerful tool to address and mitigate these issues. The implementation of BIM significantly reduces waste by improving accuracy, coordination, and decision-making during planning and construction stages (Eze et al. 2024). One notable application of BIM in waste reduction is in managing reinforcement materials, particularly rebar. A common form of material wastage in reinforced concrete construction is reinforcement/rebar trim loss, which occurs when rebar pieces are cut to fit and excess material is discarded. To tackle this, an optimization model is used that integrates a one-dimensional cutting waste minimization algorithm with BIM workflows. This model analyzes reinforced concrete structures to optimize the cutting patterns of discrete rebar elements, thereby reducing material wastage.

BIM acts as a centralized platform for integrating and communicating design information across multidisciplinary teams. By leveraging BIM's capabilities, project teams can simulate both architectural and structural requirements, compare multiple design alternatives, and rapidly implement necessary modifications to minimize wastage, especially rebar wastage. This iterative and model-based approach not only improves material efficiency but also supports cost-effective decision-making by evaluating the implications of different design choices early in the process. BIM-based optimization approach enhances the ability of planners to coordinate effectively, streamline material usage, and ultimately reduce unnecessary costs and environmental impacts associated with construction wastage. The early detection and correction of potential inefficiencies made possible by BIM ensures that projects move forward with higher precision (Porwal and Hewage 2012). BIM contributes significantly to the reduction of construction waste by addressing the root causes during design and planning stages through optimizing material use, enabling collaboration, and data-driven decision-making.

5.7 BIM and GIS Integration in the Planning Stage

The integration of BIM and GIS-based technologies is being explored for the development of tools and techniques for construction planning. Construction and maintenance of bridges often bring negative impacts to an urban environment in terms of disturbance, traffic jams and disruptions, noise, dust, and air pollution. The lack of coordination in the strategic and construction planning process is the main factor behind these negative impacts. The interoperability between BIM and GIS helps in low-disturbance construction planning. It requires knowledge of BIM, GIS, and interoperability between BIM and GIS (Sebastian et al. 2013). The use of BIM and GIS in infrastructure projects is limited because the industry foundation classes (IFC) open standards are currently targeting the building sector only. On the other hand, open GIS mainly relies on the use of geography markup language (GML)/CityGML standards. The use of BIM and GIS integration is being explored for construction planning (Youn et al. 2014), visualization of the construction process virtually before it starts (Kim et al. 2014; Song et al. 2017), indoor geo-visual analysis (Wu and Zhang 2016), etc.

Site layout planning is an essential process that has a considerable influence on safety during construction, project cost, and the time duration of a project. BIM and GIS integration is used in site layout planning to provide users with 4D visualization, placement of virtual temporary facilities, rule-based dynamic conflicts detection, route planning, hauling, and spatiotemporal analysis in a single environment (AlSaggaf and Jrade 2023).

5.8 Conclusions

Understanding the difference between construction planning, when planned individually and when planned with respect to existing facilities/utilities to have a coordinated and integrated solution is very important. Individual versus integrated construction planning is like planning in a single dimension versus n-dimensions. However, the interaction between a construction plan of an infrastructure and its surroundings is not well understood without GIS. GIS helps a construction planner in understanding the impact of surroundings on the developed construction plan. A construction plan is finalized by keeping in view the geometrical relationships and site constraints among existing and proposed infrastructure. 4D modeling of the construction process along with its surroundings in the planning stage allows a planner to visualize the effect of planning decisions and hypothesize virtually whether they will work or not, instead of mentally conceptualizing them. The use of GIS in the infrastructure construction planning stage is an area that needs future research consideration. It requires the convergence of multi-disciplinary professionals together in contrast to the current fragmented nature of the existing construction planning practices. The ever-changing priorities of sustainable development and the need to minimize environmental impacts increase complexity in the construction planning process which requires the use of GIS.

BIM has transformed construction planning by offering an integrated data-rich approach that enhances visualization, collaboration, and decision-making. Its 4D modeling capabilities link execution schedules with building components, support execution sequencing, conflict detection, site layout planning, equipment optimization, and safety management. BIM streamlines logistics and supply chain processes and enables just-in-time delivery of prefabricated components. It improves scheduling accuracy through automation, advanced simulations, and optimization algorithms. BIM-GIS integration supports construction planning by improving coordination, reducing environmental disruptions, and enabling low-disturbance strategies. Despite interoperability challenges between IFC and CityGML standards, applications include virtual construction visualization, geo-analysis, safety planning, conflict detection, and spatiotemporal analysis.

5.9 Exercises

Question 5.1: Why is planning considered a "bridge between ideas and their execution" in infrastructure projects, and what makes it a challenging task?

Question 5.2: What are the main differences between pre-project, pre-bid, pre-construction, and during-construction planning, and why is pre-construction planning considered most critical?

Question 5.3: How do knowledge-based systems contribute to construction planning, and what are their limitations regarding the role of a human planner?

Question 5.4: In what ways do 3D, 4D, and VR enhance the construction planning process, and what challenges do they address compared to knowledge-based systems?

Question 5.5: Why is the integration of BIM and GIS necessary in the construction planning stage?

Question 5.6: How does GIS support construction planning through spatial analyses such as location-based selection and area-based selection?

Question 5.7: What role do zonation maps play in assigning site-specific production rates, and how do environmental and geological factors influence productivity?

Question 5.8: How does the integration of the CPM with GIS improve construction scheduling as compared to the simple CPM?

Question 5.9: What are the key advantages of using integration of the CPM and GIS for scheduling of infrastructure projects, especially in terms of updating, visualization, and linking spatial and non-spatial data with the schedule?

Question 5.10: How does GIS support LBP and scheduling in linear infrastructure projects?

Question 5.11: Explain different types of relationships used to link 3D components with activities in the schedule to develop 4D models of the execution sequence in GIS.

Question 5.12: In what ways does 4D modeling in GIS help in detecting and resolving logical errors or missing activities in the construction schedule?

Question 5.13: Why is 4D GIS considered an effective tool for resolving workspace conflicts in large construction sites?

Question 5.14: How does GIS support the generation of dynamic workspace requirements for construction activities?

Question 5.15: Define a space conflict and explain the strategies used in GIS to resolve such conflicts.

Question 5.16: How does GIS help in developing a safety database, and what role does it play in construction safety planning?

Question 5.17: How does 4D GIS contribute to hazard identification and prevention during the construction stage?

Question 5.18: In what ways does GIS enhance construction planning visualization beyond 4D CAD, VR, and BIM?

References

Abbasi, S., Taghizade, K., and Noorzai, E. (2020). "BIM-based combination of takt time and discrete event simulation for implementing just in time in construction scheduling under constraints." *Journal of Construction Engineering and Management*, 146(12), 04020143.

Akinci, B., Fischer, M., and Kunz, J. (2002). "Automated generation of work spaces required by construction activities." *Journal of Construction Engineering and Management*, 128(4), 306–315.

Akponeware, A. O., and Adamu, Z. A. (2017). "Clash detection or clash avoidance? An investigation into coordination problems in 3D BIM." *Buildings*, 7(3), 75.

Akram, R., Thaheem, M. J., Nasir, A. R., Ali, T. H., and Khan, S. (2019). "Exploring the role of building information modeling in construction safety through science mapping." *Safety Science*, 120, 456–470.

AlSaggaf, A., and Jrade, A. (2023). "ArcSPAT: An integrated building information modeling (BIM) and geographic information system (GIS) model for site layout planning." *International Journal of Construction Management*, 23(3), 505–527.

Aram, S., Eastman, C., and Sacks, R. (2013). "Requirements for BIM platforms in the concrete reinforcement supply chain." *Automation in Construction*, 35, 1–17.

Atay, H., and Toz, G. (2011). "GIS as a planning tool for health and safety management in linear construction projects." *Journal of Earth Science and Engineering*, 1, 132–139.

Bansal, V. K. (2007). "Potential of GIS to find solutions to space related problems in construction industry." Proceedings of *World Academy of Science Engineering and Technology*, Bangkok, 26, 307–310.

Bansal, V. K. (2011a). "Use of GIS and topology in the identification and resolution of space conflicts." *Journal of Computing in Civil Engineering*, 25(2), 159–171.

Bansal, V. K. (2011b). "Application of geographic information systems in construction safety planning." *International Journal of Project Management*, 29(1), 66–77.

Bansal, V. K. (2014). "Use of geographic information systems in spatial planning: A case study of an institute campus." *Journal of Computing in Civil Engineering*, 28(4), 1–12.

Bansal, V. K., and Pal, M. (2007). "Potential of geographic information systems in building cost estimation and visualisation." *Automation in Construction*, 16(3), 311–322.

Bansal, V. K., and Pal, M. (2008). "Generating, evaluating, and visualizing construction schedule with geographic information systems." *Journal Computing Civil Engineering*, 22(4), 233–242.

Bansal, V. K., and Pal, M. (2009). "Construction schedule review in GIS with a navigable 3D animation of project activities." *International Journal of Project Management*, 27(5), 532–542.

Bansal, V. K., and Pal, M. (2011). "Construction projects scheduling using GIS tools." *International Journal of Construction Management*, 11(1), 1–18.

Bortolini, R., Formoso, C. T., and Viana, D. D. (2019). "Site logistics planning and control for engineer-to-order prefabricated building systems using BIM 4D modeling." *Automation in Construction*, 98, 248–264.

Chahrour, R., Hafeez, M. A., Ahmad, A. M., Sulieman, H. I., Dawood, H., Rodriguez-Trejo, S., Kasseme, M., Naji, K. K., and Dawood, N. (2021). "Cost-benefit analysis of BIM-enabled design clash detection and resolution." *Construction Management and Economics*, 39(1), 55–72.

Charehzehi, A., Chai, C., Md Yusof, A., Chong, H. Y., and Loo, S. C. (2017). "Building information modeling in construction conflict management." *International Journal of Engineering Business Management*, 9, 1847979017746257.

Chen, X., Zhu, Y., Chen, H., Ouyang, Y., Luo, X., and Wu, X. (2021). "BIM-based optimization of camera placement for indoor construction monitoring considering the construction schedule." *Automation in Construction*, 130, 103825.

Dashti, M. S., RezaZadeh, M., Khanzadi, M., and Taghaddos, H. (2021). "Integrated BIM-based simulation for automated time-space conflict management in construction projects." *Automation in Construction*, 132, 103957.

Dasović, B., Galić, M., and Klanšek, U. (2019). "Active BIM approach to optimize work facilities and tower crane locations on construction sites with repetitive operations." *Buildings*, 9(1), 21.

Dollner, J., and Hinrichs, K. (2000). "An object-oriented approach for integrating 3D visualization systems and GIS." *Computers and Geosciences*, 26(1), 67–76.

Eze, E. C., Aghimien, D. O., Aigbavboa, C. O., and Sofolahan, O. (2024). "Building information modelling adoption for construction waste reduction in the construction industry of a developing country." *Engineering, Construction and Architectural Management*, 31(6), 2205–2223.

Faghihi, V., Reinschmidt, K. F., and Kang, J. H. (2014). "Construction scheduling using genetic algorithm based on building information model." *Expert Systems With Applications*, 41(16), 7565–7578.

Gunes, A. E., and Kovel, J. P. (2000). "Using GIS in emergency management operations." *Journal of Urban Planning and Development*, 126(3), 136–149.

Hammad, A., Itoh, Y., and Nishido, T. (1993). "Bridge planning using GIS and expert system approach." *Journal of Computing in Civil Engineering*, 7(3), 278–295.

Hartmann, T., Van Meerveld, H., Vossebeld, N., and Adriaanse, A. (2012). "Aligning building information model tools and construction management methods." *Automation in Construction*, 22, 605–613.

Hasannejad, A., Sardrud, J. M., and Shirzadi Javid, A. A. (2023). "BIM-based clash detection improvement automatically." *International Journal of Construction Management*, 23(14), 2431–2437.

Hu, Z., Zhang, J., and Deng, Z. (2008). "Construction process simulation and safety analysis based on building information model and 4D technology." *Tsinghua Science and Technology*, 13, 266–272.

Isikdag, U., Underwood, J., and Aouad, G. (2008). "An investigation into the applicability of building information models in geospatial environment in support of site selection and fire response management process." *Advanced Engineering Informatics*, 22(4), 504–519.

Jia, N., Xie, M., and Chai, X. (2012). "Development and implementation of a GIS-based safety monitoring system for hydropower station construction." *Journal of Computing in Civil Engineering*, 26(1), 44–53.

Kenley, R., and Seppanen, O. (2010). *Location Based Management for Construction Planning, Scheduling and Control*. London: Spon Press.

Kim, J. E., Kang, T. W., and Hong, C. H. (2014). "Development of the spatial indexing method for the effective visualization of BIM data based on GIS." *Journal of the Korea Academia-Industrial Cooperation Society*, 15(8), 5333–5341.

Koo, B., and Fischer, M. (2000). "Feasibility study of 4D CAD in commercial construction." *Journal Construction Engineering and Management*, 126(4), 251–260.

Kumar, S. S., and Cheng, J. C. (2015). "A BIM-based automated site layout planning framework for congested construction sites." *Automation in Construction*, 59, 24–37.
Le, P. L., Dao, T. M., and Chaabane, A. (2019). "BIM-based framework for temporary facility layout planning in construction site: A hybrid approach." *Construction Innovation*, 19(3), 424–464.
Liu, H., Al-Hussein, M., and Lu, M. (2015). "BIM-based integrated approach for detailed construction scheduling under resource constraints." *Automation in Construction*, 53, 29–43.
Liu, H., Lei, Z., Li, H. X., and Al-Hussein, M. (2014). "An automatic scheduling approach: Building information modeling-based onsite scheduling for panelized construction." *Construction Research Congress 2014: Construction in a Global Network*, 1666–1675.
Melzner, J., Zhang, S., Teizer, J., and Bargstädt, H. J. (2013). "A case study on automated safety compliance checking to assist fall protection design and planning in building information models." *Construction Management and Economics*, 31(6), 661–674.
Mirzaei, A., Nasirzadeh, F., Parchami Jalal, M., and Zamani, Y. (2018). "4D-BIM dynamic time–space conflict detection and quantification system for building construction projects." *Journal of Construction Engineering and Management*, 144(7), 04018056.
Ocheoha, I. A., and Moselhi, O. (2018). "A BIM-based supply chain integration for prefabrication and modularization." *Modular and Offsite Construction (MOC) Summit Proceedings*.
OpenAI (2025). ChatGPT (Version 5.1) [Large language model]. https://chat.openai.com/
Parsamehr, M., Dodanwala, T. C., Perera, P., and Ruparathna, R. (2023). "Building information modeling (BIM)-based model checking to ensure occupant safety in institutional buildings." *Innovative Infrastructure Solutions*, 8(6), 174.
Poku, S. E., and Arditi, D. (2006). "Construction scheduling and process control using geographic information systems." *Journal of Computing in Civil Engineering*, 20(5), 351–360.
Porwal, A., and Hewage, K. N. (2012). "Building information modeling–based analysis to minimize waste rate of structural reinforcement." *Journal of Construction Engineering and Management*, 138(8), 943–954.
Sebastian, R., Böhms, M., and van den Helm, P. (2013). "BIM and GIS for low-disturbance construction." In: N. Dawood and M. Kassem (Eds.), Proceedings of 13th International Conference on *Construction Applications of Virtual Reality*, 30-31 October 2013, London, UK, 469–479.
Seo, J., and Kang, S. (2006). "Geographic information system based roadway construction planning." *Canadian Journal of Civil Engineering*, 3(5), 508–520, 2006.
Sharma, S., and Bansal, V. K. (2018). "Location-based planning and scheduling of highway construction projects in hilly terrain using GIS." *Canadian Journal of Civil Engineering*, 45(7), 570–582.
Sharma, S., Bansal, V. K., and Parti, R. (2014). "Current state of highway projects planning and scheduling." *International Journal of Information Technology Project Management*, 5(4), 50–67.
Sharma, S., Bansal, V. K., and Parti, R. (2016). "Location-based highway construction planning using geographic information systems." *International Journal of Construction Management*, 16(4), 299–314.
Song, Y., Wang, X., Tan, Y., Wu, P., Sutrisna, M., Cheng, J. C. P., and Hampson, K. (2017). "Trends and opportunities of BIM-GIS integration in the architecture, engineering and construction industry: A review from A spatio-temporal statistical perspective." *ISPRS International Journal of Geo-Information*, 6(12), 397.
Sonmez, R., and Uysal, F. (2008). "Geographic information system-based visualization system for planning and monitoring of repetitive construction." *Canadian Journal of Civil Engineering*, 35, 1342–1346.

Tauscher, E., Smarsly, K., König, M., and Beucke, K. (2014). "Automated generation of construction sequences using building information models." *Computing in Civil and Building Engineering*, 745–752.

Underwood, J., and Isikdag, U. (2010). *Handbook of Research on Building Information Modeling and Construction Informatics: Concepts and Technologies*. Hershey, NY: Information Science Reference.

Wang, W. C., Weng, S. W., Wang, S. H., and Chen, C. Y. (2014). "Integrating building information models with construction process simulations for project scheduling support." *Automation in Construction*, 37, 68–80.

Wong, N. H., and Jusuf, S. K. (2007). "GIS-Based and computer simulation evaluation on campus master plan." Proceedings 2nd PALENC Conference and 28th AIVC *Conference on Building Low Energy Cooling and Advanced Ventilation Technologies in the 21st Century*, 1121–1126.

Wu, B., and Zhang, S. (2016). "Integration of GIS and BIM for indoor geovisual analytics." *International Archives of the Photogrammetry, Remote Sensing and Spatial Information Sciences*, Volume XLI-B2, 2016 XXIII ISPRS Congress, 12–19 July 2016, Prague, Czech Republic.

Youn, J., Kang, T. W., Choi, H., and Kim, S. (2014). "Derivation of BIM/GIS platform application scenarios and definition of specific functions in construction planning and design." *Journal of the Korea Academia-Industrial Cooperation Society*, 15(12), 7340–7349.

Zhang, S., Teizer, J., Lee, J. K., Eastman, C. M., and Venugopal, M. (2013). "Building information modeling (BIM) and safety: Automatic safety checking of construction models and schedules." *Automation in Construction*, 29, 183–195.

Zhang, S., Teizer, J., Pradhananga, N., and Eastman, C. M. (2015). "Workforce location tracking to model, visualize and analyze workspace requirements in building information models for construction safety planning." *Automation in Construction*, 60, 74–86.

Zhong, D., Li, J., Zhu, H., and Song, L. (2004). "Geographic information system based visual simulation methodology and its application in concrete dam construction processes." *Journal of Construction Engineering and Management*, 130(5), 742–750.

Zlatanova, S., and Prosperi, D. (2005). *Large-Scale 3D Data Integration: Challenges and Opportunities*. Boca Raton, FL: CRC Press.

Chapter 6

The Construction Stage

6.1 Learning Objectives

After the completion of this chapter, readers will be able to:

- understand the need and use of GIS in the construction stage,
- understand the use of GIS as an alternative for construction site layout planning,
- understand the use of GIS for excavation planning, tracking site dynamics, and reducing construction waste, and
- understand the application areas of BIM and the integration of BIM and GIS in the construction stage.

6.2 Introduction

Productivity during the construction stage is of significant importance in the AEC industry. To increase productivity, the main emphasis is on labor and equipment productivity, which generally overlooks other potential areas that may yield significant improvement in productivity during the construction stage. The macro-level planning during the pre-construction stage and the micro-level planning during the construction stage are viable areas to improve productivity during the construction stage. The effective use of software tools and techniques during the construction stage has a significant impact on the infrastructure construction project duration, cost, and quality. The use of GIS, as a software tool, provides a virtual environment to support various types of geospatial analyses during the construction stage. The construction stage of infrastructure projects has a high level of complexity. Therefore, it is planned and scheduled with respect to the daily life activities of the people involved, people being affected by the construction activities, and the surroundings of an infrastructure being constructed. As discussed in the previous chapters, GIS integrates information which are used during the construction stage. The information, such as execution schedule, working drawings, specifications, site layouts, and other project documents on a single GIS platform, which are used during the construction stage, allows a better

DOI: 10.1201/9781003759348-6

understanding of the project execution. GIS supports in optimizing infrastructure programs of distributed infrastructure construction sites (Hegazy 2005). Several application areas of GIS during the construction stage are discussed in the subsequent sections to keep the AEC industry professionals informed about the use of GIS. This chapter, in the following sections, covers the application areas of GIS, BIM, and integration of BIM and GIS in the construction stage of infrastructure projects.

6.3 GIS in the Construction Stage

Contractor selection, workforce planning, materials trading, and site layout design critically affect the construction stage. GIS enhances contractor selection, e-commerce activities, site layout planning, decision-making, and sustainable infrastructure development through the use of optimum resources. GIS minimizes materials losses and plans TFs with safety considerations and topography in view. It optimizes excavation logistics, equipment placement, progress tracking, and waste reduction. The following sections provide details:

6.3.1 *Screening of Contractors*

Generally, a contractor to do a particular work is selected on the basis of the rate quoted in his bid, experience, and reputation. A large volume of qualitative, subjective, and imprecise information are typically involved in the prequalification process of a contractor to do a particular type of work. Numerous models are available to rationally structure and formalize the contractor selection process for doing the particular work. However, no attention is given to the geographical aspect of the contractor selection evaluation problem to do work on the particular geographic location. A geographical location has a great influence on the capability to do the work and the output of the work of a contractor. GIS is used to consider the geographical aspect for screening contractors in order to determine their competence to complete infrastructure construction work at the particular geographical location (Jeljeli et al. 1993).

6.3.2 *Human Workforce Planning*

The infrastructure construction market includes many players such as employers, organizations, the government, and trade unions. All play distinctive and varying roles in present and future labor market planning. Infrastructure construction projects involve a large volume of human resources; hence, planning human workforce needs for the present and future is always crucial. Human workforce planning must be as accurate as possible; it must take into consideration the shortages of workforce at a particular location that may occur in the present or future. Various planning methods used for labor market planning possess several limitations. GIS provides an opportunity to better understand geographical influences on labor market issues and facilitates policies for the future to be tailored to suit the individual needs of a geographic location. GIS allows the integration and manipulation of diverse datasets to give a clearer understanding of the geographical trends of the labor market for future planning (Anumba et al. 2008).

6.3.3 *Construction Materials Trading*

The web-based technology has made it much easier for the AEC industry practitioners to maintain materials trading information and share it over the internet regardless of their physical locations. Construction materials trading involves suppliers, buyers, and agents as three major players. Agents help suppliers and buyers to complete trading/transactions. The suppliers and buyers are required in all trading activities, while agents exist only in certain trading situations. The links between suppliers and buyers are organized in three ways: (a) suppliers and buyers form direct connections, (b) suppliers and buyers form connections through or with the help of agents, and (c) suppliers and buyers form connections through web-based electronic markets. The web-based GIS model used for e-commerce business in the AEC industry is based on the type (c) trading situation (Li et al. 2003). The e-commerce system in GIS is used for online orders and offline delivery of different infrastructure construction materials.

The web-based electronic market provides a platform for suppliers to provide online information about their products. Buyers easily search and compare products of different suppliers through an online system and contact the suppliers directly or buyers may also invite agents to undertake certain tasks required in order to complete a trading/transaction. Type (c) trading has the most flexibility and functionality to support all infrastructure trading situations encountered in construction materials trading, and does not support bargaining and bidding trading situations.

In all kinds of construction activities, construction material transportation costs are involved; consequently, the cost of transporting materials is a critical consideration in many e-commerce trading. The web-based GIS provides an ideal solution to find the possibility of delivering infrastructure construction materials to a particular location and calculates the costs of transportation. However, sometimes the cost of transportation of construction materials does not only depend on the distance but involves some other variables such as the locations of local distributors that can reduce transportation costs or by combining shipments to various buyers in the same geographical area (Li et al. 2003).

6.3.4 *Site Layout Planning*

In the construction of infrastructure projects, the AEC industry professionals focus on productivity in terms of cost, quality, and time, neglecting the site layout planning. The AEC industry professionals believe that focusing on site layout planning means delaying planning progress and increasing project costs. The construction site layout planning is sometimes done based on the experience of a site engineer. Site layout planning affects the operations of various temporary facilities (TFs), productivity, and safety during the construction stage. Site layout planning is not an easy task due to the involvement of various types of resources involved, the need for materials storage yards, interlinked constraints, complex topography, etc. The construction site layout planning, where site space is limited, needs to be done carefully. On the other hand, planning for large construction sites having abundant availability of site space, proper positioning of materials and TFs with respect to each other has a huge impact on site operations, materials handling, and transportation costs. In every situation, construction site layout planning is crucial and affects production rates and safety during the construction stage.

The construction site layout planning is broadly divided into three categories: static layout planning, dynamic layout planning, and space scheduling. A site layout is termed as static when only one site layout is used for the entire construction stage. However, the changing nature of a construction project and its impact on a site make the construction site layout outdated only when little progress is made on a site. The requirements of a site change considerably with time throughout the construction stage. As construction proceeds, more and more space is occupied by the permanent facilities/infrastructure, leaving less space to position materials and support TFs. The process of developing a site layout that changes over time to meet the requirements of a site as construction progresses is termed as dynamic site layout. Dynamic site layout planning allocates space for materials storage and TFs for a particular time duration on predetermined locations at a site (Cheng and O'Connor 1996). Allocating site space as per the requirements of components of an infrastructure being constructed over time, as governed by the construction schedule, is termed as space scheduling. Activity-level space scheduling involves the allocation of site space over time so as to complete the construction activities in a time-bound manner.

The construction site layout planning is done carefully, keeping in view the requirements of construction activities to be executed and constraints in terms of shape, size, topography, traffic, and other restrictions. A poorly planned construction site causes accidents resulting from falls of materials from height and collisions between workers, plants, or equipment. Whereas a well-planned construction site layout enables safe, smooth, and efficient construction operations while reducing the space required for TFs.

One of the various factors that control the success of a project is the allocation of space to various site elements during construction to develop a site layout. Site layout planning is normally accomplished by the AEC industry professionals through trial-and-error methods. It is generally the problem of identifying the numbers and sizes of TFs to be placed, identifying constraints between TFs, determining relative positions of TFs that satisfy constraints to allow them to function properly, deciding materials storage locations and their sizes, deciding the distances between the materials storage locations and corresponding activities, etc. It is ideally a dynamic problem because of so many changes take place during the construction stage;

therefore, updating a site layout is a prerequisite for a productive construction site (Hegazy and Elbeltagi 1999).

Various techniques are used for developing site layouts for a construction project. The problem-solving techniques for site layout planning range from knowledge-based heuristic methods to mathematical models. The knowledge-based heuristic methods provide many good solutions, whereas mathematical models provide more accurate and optimum solutions. However, the widely used techniques for construction site layout planning are still heuristic. The widely used techniques reported for site planning are CAD-based, genetic algorithms-based, expert system-based, virtual reality-based, fuzzy sets theory-based, neural networks-based, spreadsheet-based, and building information modeling-based (Kumar and Bansal 2019). GIS improves construction site layout planning by integrating location-based information and attributes into a single environment. The site layout planning in GIS is performed through the identification of locations suitable for locating TFs or storing materials in order to minimize leads, construction space conflicts, and improve site productivity. GIS helps in developing an automated site layout system for identifying suitable locations for materials storage and TFs by integrating a database management system and geospatial analyses.

6.3.4.1 Safety Considerations in Site Layout Planning

Studies indicate that a significant amount of time is lost due to work-related health and safety problems on construction sites. The AEC industry around the world is implementing health and safety management systems to provide a safe working environment on construction sites. The need for health and safety-related systems among construction professionals has increased due to the involvement of risky and unhealthy operations during the construction stage of infrastructure facilities. The risky and unhealthy operations result in many human tragedies, discourage workers, disrupt construction, delay progress, affect cost, decrease productivity, and adversely affect the reputation. Ignoring health and safety means more accidents, invitations to litigation, an increase in compensation of workers, more insurance premiums, etc.

Construction site accidents are avoided by giving ample consideration to health and safety-related issues during the early stage of site layout planning. Accordingly, several causes of construction site accidents and health hazards, such as falls of humans, falling objects on humans, site transportation, and exposure to hazardous substances, are controlled through creating an efficient site layout plan. Therefore, safe site layout planning is the best way for construction professionals to address several health and safety-related issues. The health and safety-related information are fragmented; construction professionals have poor knowledge about safety-related practices. GIS is being used in developing safe site layouts (Kumar and Bansal 2021). Proper safety zones around construction areas planned in a site layout prevent harm from falling objects. These zones protect workers and entities from falling objects. The safety zones around a component under construction are considered in the site layout planning done using GIS. The site elements are located based on safety code provisions in GIS-based site layout planning. The inclusion of safety considerations in the site layout planning using GIS improves construction safety by connecting safety issues more closely to the execution plan.

6.3.4.2 *Site Layout Elements and Their Area Requirements*

Most of the construction sites are demonstrated as a set of three interrelated elements, known as site layout elements: site entities, TFs, and site constraints. All permanent elements on a construction site whose locations govern or are used to fix the locations of new elements on a site are called site entities. For example, the new infrastructure facility to be constructed, existing facilities/utilities, land use, overhead power supply lines, roads, etc., are site entities. TFs are movable or fixed ones that occupy some space on a construction site. Identifying the required numbers and determining the area requirements of TFs for constructing an infrastructure are difficult decisions. These require thorough consideration of project conditions and code provisions. In some projects, TFs affecting construction site safety are also identified and their impacts are quantified. TFs such as temporary site offices, labor rest areas, water storage tanks, rebar fabrication, storage yards, laydown yards, etc., serve the construction of an infrastructure facility. The presence of trees, water ponds, steep slopes, site boundary, runoff patterns, etc., are known as site constraints that require due consideration in locating TFs and developing a site layout. When site constraints are satisfied, then optimum locations for TFs are decided. The constraints also exist between TFs and site entities or between one TF with other TFs, depending upon an infrastructure construction project. GIS helps in the modeling of locations and area requirements among site elements (Kumar and Bansal 2016).

6.3.4.3 *Materials Layout Planning*

The materials layout planning on a construction site is crucial in planning the execution of infrastructure construction projects. A poor materials layout at a site causes site congestion and inaccessibility to certain materials which sometimes leads to project delays and cost overruns. The materials estimate is integrated with the execution schedule to generate dynamic materials requirements in GIS. Based upon the information regarding the quantities of various types of materials required to be stored and locations available on a site, GIS is used to identify suitable locations on a site to store construction materials. GIS may be designed to pass on the information dynamically to a site for material layout planning (Cheng and Yang 2001). The rules of thumb and experience are used to decide the size of materials storage areas and the placement of materials to develop a materials layout plan.

GIS is also used to develop materials accessibility grades to quantify materials accessibility during the project execution. A materials layout evaluation model in GIS is used based on materials accessibility grades (Su et al. 2012). GIS platform automates tasks like: reading user's inputs, generating lists of materials in demand and supply, calculating grades for pick-up convenience, delivery convenience, shortest path, and calculating overall materials accessibility grades for developing site layout plans. The dynamic materials layout is obtained by linking the execution schedule with the materials requirement and finally linking the materials schedule with the materials supply. GIS helps in identifying space conflicts, generating conflict-free execution plans, and calculating materials accessibility grades for all conflict-free execution plans.

In hill areas, improper placement of materials, particularly coarse and fine aggregates, leads to huge wastage due to washing away along with floodwater. Materials are kept away or at a safe distance from floodwater. GIS helps in the identification of locations to store materials in order to avoid being washed away along with floodwater.

6.3.4.4 TFs Layout Planning

TFs layout planning identifies locations within the available site space for locating various TFs, keeping their shapes and sizes into consideration to support various operations during the construction stage. TFs serve construction operations but are not considered as a physical part of any infrastructure facility to be developed. Primarily, TFs are required to support construction operations and provide services for workers on a site (Cheng and O'Connor 1994). Examples of the widely used TFs are: materials stores, fabrication yards, lay-down areas, parking lots, site offices, warehouses, etc. All infrastructure construction projects require a considerable number of TFs for their execution. The GIS-based procedure, as shown in Figure 6.1, is used to identify locations for locating TFs on a construction site. In general, the procedure has four steps: *development of input layers*, *modeling requirements*, *analyzing space* to locate TFs on a construction site, and *development of a dynamic layout*. Each step is discussed in detail in the subsequent sections:

Development of input layers: Input layers are the layers representing site layout elements. The input layers graphically represent the spatial information about the site layout elements in GIS. The construction professionals decide on layout elements that affect a site layout. The input layer corresponding to each element is developed to locate the required TFs on a site. For example, layers representing existing roads, existing facilities/utilities, restricted areas, safety zones, buffer zones around runoff, etc., are developed. A few layers are also used as input to develop additional layers like slope from DEM, distance from the boundary, distance from electric line, distance from existing infrastructure, distance from hazardous locations, etc.

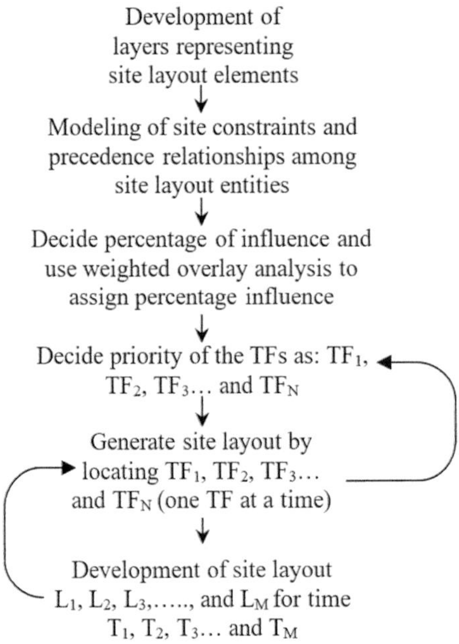

Figure 6.1 GIS-based procedure to identify locations for TFs on a site

Modeling requirements: Important aspects considered in locating TFs on a construction site are site layout elements, restricted areas, safety zones, site topography, etc. Among them, the modeling of site constraints and interrelationships among site entities are important to suitably locate TFs. An area on a construction site prone to injury is restricted from being allocated to TFs. A buffer of a certain width is marked as a restricted area around such hazardous locations to avoid the chances of any accident from a safety point of view. TFs are located on a construction site by providing appropriate distances between them. Safety zones indicate extra space required in addition to the physical area of an infrastructure component under construction. This space acts as a buffer to protect any person who might be injured by the fall of materials, tools, or equipment being raised or lowered. In order to place construction materials at suitable locations, an understanding of topography is also significant. Modeling of topography in GIS helps in reducing the wastage of materials like sand, gravel, damage to sensitive electrical equipment, etc. The surface model is developed in GIS and utilized for identifying natural streams of water, around streams buffers are created for the purpose of excluding this area while locating TFs.

Analyzing space: Numerical values in the input layers are in different ranges, making it impossible to use these layers in the present form to get a meaningful solution for locating TFs. To use the input layers, their numerical values are converted on a common measurement scale of 1 to 10. The common scale helps in weighted overlay analysis in GIS. Higher values are given to the attributes that are more suitable locations for TFs and lower values to the least suitable locations. For example, the highest value is assigned to the horizontal surface and the lowest value is given to the steeply sloped area in the input layer representing the site slope for locating TFs. GIS-based weighted overlay analysis is used to assign a percentage of influence to the re-classed layers. The percentage of influence is variable with the choices of construction professionals involved in identifying space to locate TFs. In case all factors corresponding to every layer are equally important, an equal percentage of influence is assigned to each input layer. Generally, different layers are assigned different percentages of influence; higher percentages of influence are assigned to layers having more influence in locating TFs.

Development of dynamic layout: Output map of the last step is used as the input in this step. The GIS-based procedure helps in locating TFs one by one as depicted in Figure 6.1. The first output is a map showing the location of the first TF, that is, TF_1. Each TF located earlier and the input map of the last step becomes the input for identifying the location of a subsequent TF. For example, the input for locating the second TF (TF_2) is the input of TF_1 and the location of TF_1. And for the third TF (TF_3) the input becomes inputs of TF_2 and location of TF_2 and so on. The input statement for locating TFs is generalized as input for location TF_N is equal to the input of TF_{N-1}+location of TF_{N-1}, where N is the number of TF required to support construction operations. The priority order of locating TFs depends upon the choices of construction professionals who are involved in locating all TFs. The layout L_1 represents the locations of all TFs required at time T_1. Similarly, layouts L_2, L_3, …., and L_M at times T_2, T_3 …, and T_M are developed, where M is the number of layouts to be developed.

6.3.5 *Planning Excavation Works*

Earthwork is a significant task in large infrastructure construction projects like highways, railways, etc., as it consumes major portions of both project cost and time. The studies on highway or railway construction using GIS are focused on earthwork planning only (Kang and Seo 2013). GIS is used to optimize earthmoving involved in highway construction projects (Moselhi and Hassanien 2003). GIS is used to generate a digital terrain model of site topography and soil profiles for developing an optimum earthmoving plan. Variations in the type of soil in cut sections, compaction level, groundwater levels, multiple landfills and borrow pits, and the presence of transverse obstructions such as rivers, valleys, etc., along a project length are considered in GIS for optimizing the earthwork. Genetic algorithms, GPS, and GIS are used to optimize crew formations during earthmoving operations in infrastructure construction projects (Moselhi and Alshibani 2007). The following subsections describe the use of GIS in earthwork cut and fill volumes calculation and moving earthwork:

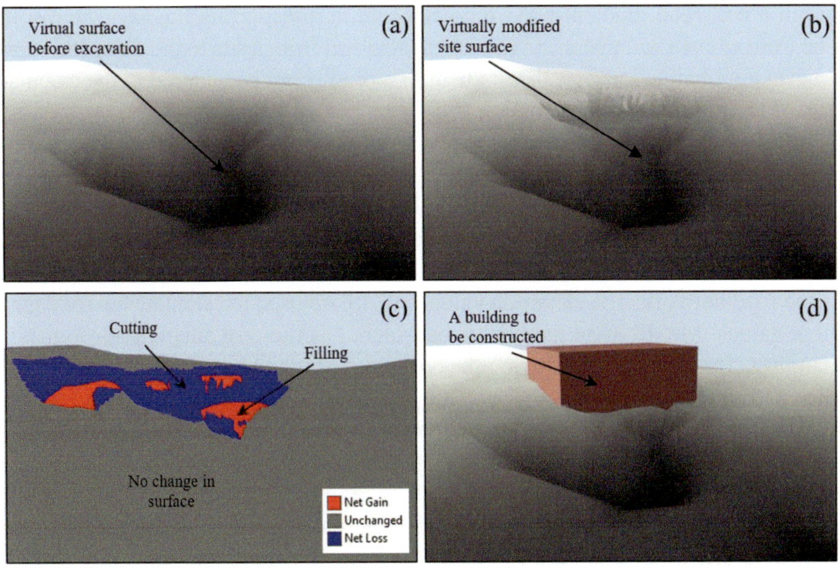

Figure 6.2 Earthwork planning, (a) virtual surface of a construction site before excavation, (b) virtually modified surface as per the requirements, (c) output depicts where the surface is cut, filled, and remains unchanged, and (d) block model of the building placed on its location

6.3.5.1 *Cut and Fill Volumes Calculation*

Earthwork planning is a valuable asset in GIS, original (natural) and final (after required topographical changes) virtual surface models, as shown in Figure 6.2(a) and (b), are the inputs for cut and fill volumes calculations. The portions on a virtual surface where the soil is to be excavated, filled, or where the surface remains unchanged are identified in GIS as shown in Figure 5.3(c). GIS helps in calculating the volumes of soil that is cut and is dumped on the locations identified for filling (Bansal 2014). Once on-site cut and fill is over, excavated soil is planned to be transported to the dumping locations. The virtual surface model is used in GIS to identify locations near a construction site having elevation lower than its surroundings as potential dumping locations.

6.3.5.2 *Moving Earthwork*

GIS is used to develop an optimum earthmoving plan (Hassanein 2002). Variations in the types of soil and presence of the transverse obstructions such as rivers, valleys, etc., along the road are also considered in optimizing the earthmoving plan in GIS. In general, operations like loading, transportation, and unloading/dumping of excavated soil are planned by considering the construction site as a leveled surface.

In hill areas, road gradient plays a crucial role in transporting the excavated soil. The true output of dumpers carrying excavated soil from a site to the dumping locations is not obtained in case the gradient of a road is not considered in the calculation of travel time or optimum path.

In hill areas, the construction involves the movement of excavated soil, machinery, and equipment/materials, hence, road gradient must be considered in site logistics planning also. To consider road gradients in site logistics planning, the centerline of a road network is draped over the virtual surface to convert a 2D road network into 3D. It helps in the calculation of elevation values along the centerline of the road network. The elevation values in GIS are used to calculate the road gradient values. In GIS, road gradient is considered as the cost attribute for logistics planning. This helps in prioritizing the dumping locations based on the least resistance path from a construction site to dumping locations. GIS helps in finding the least resistance path between a dumping location and an infrastructure construction site. This is not the shortest path between two locations but the path having the least road gradient (Bansal 2023).

GIS is used for the selection of an optimum haul route using the topographic map of a site (Seo and Kang 2006). The cost is calculated from the sum of rolling, grade, and distance resistance offered by each route to hauling equipment. GIS is used to determine the optimum layout of haul routes in large earthmoving infrastructure projects (Kang and Seo 2013). The factors considered to calculate the impedance attribute of haul routes are land use, truck speed, topography, etc. GIS is also used for carrying out the least cost path analysis between cut and fill areas. GIS is used to identify all areas of internal drainage having elevation lower than its surroundings. The depressions are problematic because water that flows into them cannot come out. These are the suitable locations for dumping excavated soil. The raster cells with a high flow accumulation are areas of concentrated flow used to identify stream channels. The raster cells with a flow accumulation of zero are local topographic highs used to identify ridges.

6.3.6 Site Drainage Planning

Infrastructure construction sites located in low-lying areas get flooded with water because of high rainfall. The rainwater comes through the steepest paths downhill to the low-lying areas. Such situations, sometimes, are not predicted while developing the site layout or work execution plan. To continue construction activities in such situations, site drainage is essential. The virtual surface model in GIS is utilized to understand the natural flow of rainwater and how changes in the topography, because of construction activities affect the flow pattern. The runoff pattern of the surroundings of an infrastructure under consideration is developed using the surface model. Flow directions of the rainwater toward the infrastructure construction sites are identified. Figure 6.3 shows the natural flow pattern and profile along line *xy* of an infrastructure construction

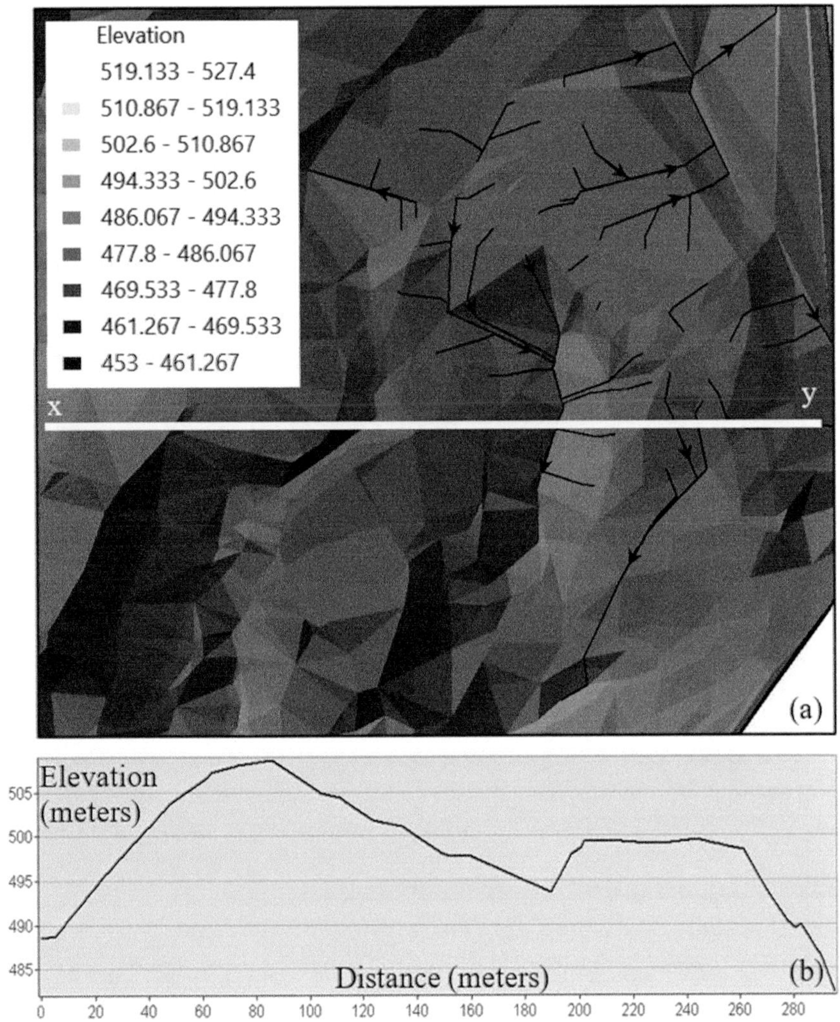

Figure 6.3 Natural flow pattern and profile along line *xy* without disturbing site topography

site without disturbing site topography. Figure 6.4 shows changes in the flow pattern and profile along the *xy* line due to the modifications of site topography to construct an infrastructure facility. The change in flow pattern is helpful in understanding the inflow and outflow of rainwater for draining a site and future drainage planning (Bansal 2017). It is clear from the figures that to continue infrastructure construction activities, the development of a site drainage plan is essential in hill areas.

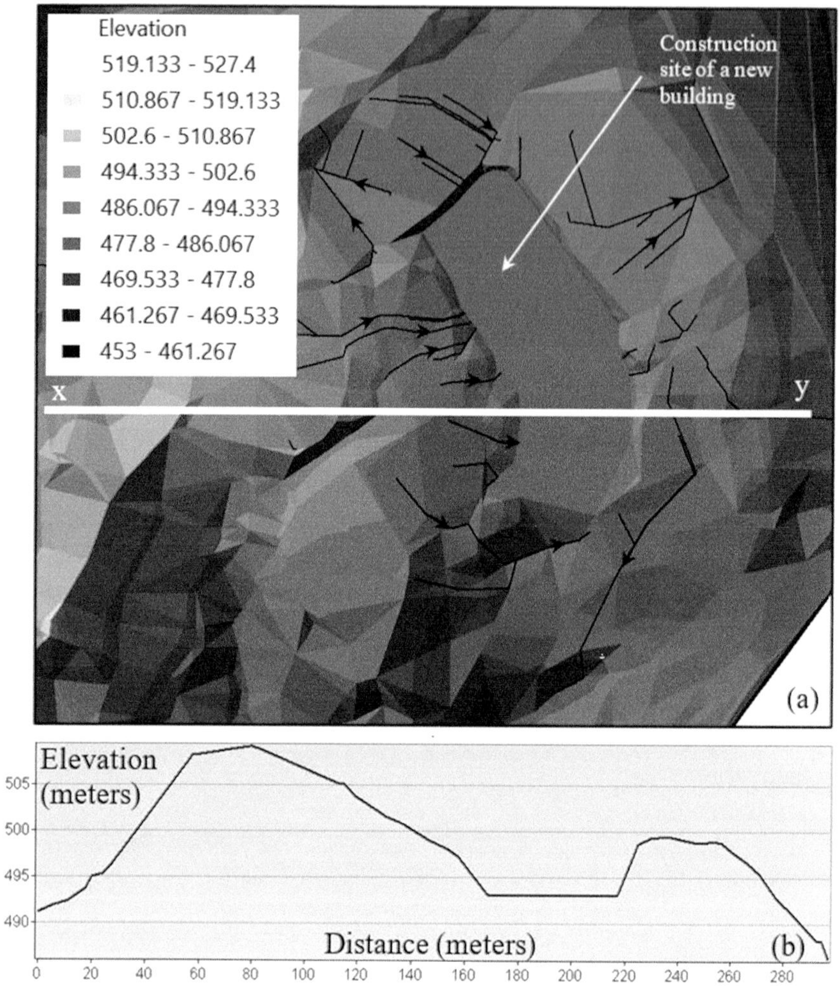

Figure 6.4 Changes in the flow pattern and profile along *xy* line due to the modifications in site topography to construct an infrastructure

6.3.7 Identification of Locations for Equipment

GIS helps in identifying the locations for setting up or positioning equipment on infrastructure construction sites. 3D visualization in GIS helps in identifying conflicts among heavy equipment like cranes on a site. The power line layout on a site, site constraints, and proximity of existing facilities/utilities that affect an equipment location are considered in the process of identification of suitable locations for equipment on sites. GIS is also used in the selection of particular equipment under specific work conditions on a construction site. In response to mitigate chaotic disaster situations, well-organized response operations are required during infrastructure rehabilitation. Efficient allocation of resources, such as equipment, is a critical task during rehabilitation. GIS is used to facilitate equipment allocation in response to disasters during rehabilitation (Chen et al. 2011).

6.3.8 Tracking Site Dynamics and Progress Control

Traditional scheduling and progress control techniques, such as bar charts and networks, fail to provide information pertaining to the spatial aspects of an infrastructure construction project. GIS is used to develop a system to represent construction progress in terms of construction graphics synchronized with the execution schedule (Poku and Arditi 2006). The design and scheduling information (including percent complete information) are plugged into GIS which help in effectively communicating the schedule/progress information to the parties involved in the project execution.

GIS is used to develop an automated schedule monitoring system to control the erection process for precast infrastructure construction projects. GIS assists construction managers in controlling and monitoring the lifting process of precast components. The structural elements are prefabricated in a manufacturing plant and are transported to a construction site for installation. The erection of prefabricated structural components is considered a major critical activity for precast construction (Cheng and Chen 2002). The schedule for prefabrication and transportation of structural elements to a site is developed based on their installation schedule in GIS. The barcode system combined with the wireless radio technology is used to collect and transmit the infrastructure construction site information to a central controlling place (Andoh et al. 2012). This integration is achieved by using an automated barcode identification system in three phases (design, manufacturing, and use) of the precast construction project. The use of GIS improves real-time schedule monitoring systems, construction processes, and enhancing efficiency.

6.3.9 Reducing Construction Wastage

GIS and GPS technologies together are used to reduce construction waste at sites (Li et al. 2005). The steps involved are automatic data capturing, barcode for construction materials and onsite equipment management, and integrated GPS and GIS technologies based on a wide area network. The controls at headquarters and construction sites get real-time information to monitor cargo on a road to various sites so as to reduce waste generation on various sites. The site control room uses accurate arrival time of construction materials and onsite equipment management to deal with limitations of materials and equipment storage on various sites and to finish construction projects in time with minimum generation of construction waste. GPS and GIS integrated solutions improve construction efficiency by increasing the effective working hours of construction equipment, reducing construction time duration, and the costs of the workforce. The initial investment in hardware and software is justified by the fact that an infrastructure construction company can utilize the developed system in other construction projects later.

6.4 BIM in the Construction Stage

BIM serves as a dynamic platform that enables the retrieval, generation, and integration of critical project data, facilitating cohesive and collaborative work across the construction stage. At its core, BIM provides a centralized environment where accurate and consistent digital information are created, shared, and reused by stakeholders. During the construction stage, BIM's multi-dimensional modeling capabilities allow for the representation of continuously evolving site conditions, the application of decision-making rules, and the evaluation of individual components for compliance with quality standards (Ding et al. 2014). These capabilities ensure that informed decisions are made in real-time, improving construction efficiency and reducing costly errors. BIM also allows for the incorporation of key performance parameters such as safety management during construction, quality control, and carbon emissions tracking. This integration provides project teams with access to actionable information, enhancing the decision-making process and ensuring that environmental and regulatory requirements are met effectively.

To advance these capabilities even further, the integration of BIM with emerging technologies such as artificial intelligence is revolutionizing the AEC industry. The use of standardized data, machine learning algorithms, natural language processing, virtual assistants, optimization techniques, and cloud-based platforms contributes to the transformation of construction processes (OpenAI 2025). These innovations support real-time analytics, predictive modeling, automated quality checks, and intelligent decision support systems, resulting in improved productivity, reduced risk, and enhanced project outcomes.

In the design and construction of underground transportation infrastructure, BIM is used to develop a 3D subsoil model that integrates geological and geotechnical data. Once a tunnel's BIM model is defined, including the structural and subsoil parameters, it is exported to a finite element model for deformation analysis (Fabozzi et al. 2021). Additionally, 4D-BIM is employed to manage and simulate the tunnel excavation sequence over the construction schedule. This geotechnical integration enhances the reliability and usefulness of the model not only during design but especially throughout construction and long-term infrastructure management, particularly in complex urban environments where informed decision-making is critical.

With the rapid urbanization of cities, the demand for underground space has led to the widespread development of subterranean infrastructure such as deep excavations, tunnels, and basements. These structures are common in older or densely populated urban environments where space is limited. However, constructing deep excavations in such areas introduces significant risks. The process can adversely affect ground stability and potentially cause structural damage to adjacent buildings. In severe cases, accidents during excavation can result in property loss, structural failures, and even human casualties. Given these risks, it is essential to thoroughly evaluate and continuously monitor the environmental and structural impacts of deep excavation activities throughout their construction process. Key concerns typically include ground movement, soil settlement, water table fluctuations, and stress on retaining structures. To address these, monitoring systems are employed to track various parameters during critical phases such as the installation of retaining walls, excavation of soil, and groundwater pumping operations. These activities generate vast amounts of data, ranging from geotechnical measurements and structural deformation reports to real-time environmental sensor outputs. Managing and making sense of this complex and voluminous information manually or using traditional methods becomes increasingly inefficient and error-prone as the project progresses. This is where BIM plays a transformative role.

BIM-based monitoring systems provide an integrated digital platform for managing and visualizing excavation-related data in real-time. By consolidating 3D models of the project with geotechnical data, environmental assessments, and ongoing monitoring results, BIM allows construction teams to analyze and interpret complex datasets more effectively (Wu et al. 2015). These digital models typically include detailed representations of the excavation geometry, structural support systems, adjacent buildings, and soil layers, enabling a holistic understanding of the excavation environment. Moreover, the integration of BIM with real-time monitoring instruments enhances the ability to detect anomalies early. For instance, data from instruments measuring ground surface settlement, lateral wall movements, and groundwater levels can be linked directly to the 3D excavation model. This visual and spatially-aware presentation of data helps project teams quickly identify potential safety issues or deviations from expected conditions, allowing for immediate intervention before problems escalate (Lu et al. 2012). Such proactive monitoring not only improves risk management but also ensures compliance with regulatory requirements and environmental standards. It enhances communication among stakeholders by providing a common visual language and centralized data source, fostering more coordinated decision-making. The integration of BIM in urban deep excavation projects significantly improves the management of environmental risks and structural safety. By merging real-time data with intelligent models, BIM enhances situational awareness, supports accurate environmental impact assessments, and contributes to the successful and safe delivery of complex underground construction projects.

BIM contributes across a wide range of construction-related tasks. These include ensuring occupational health and safety compliance, monitoring excavation progress, tracking material usage, managing quality assurance, documenting construction activities, and even supporting delay claim calculations through precise progress data and visual records. The following sections will delve deeper into the diverse and transformative applications of BIM during the construction stage, highlighting how this technology continues to redefine the delivery and management of infrastructure construction projects:

6.4.1 *Construction Progress Monitoring*

BIM plays a critical role in enhancing communication, coordination, and site management during the construction stage. Its application supports more effective control over materials and monitoring of construction progress (Nadeem et al. 2017). One of the most impactful features of BIM is its visualization capabilities, which allow engineers, site managers, and planners to better understand the actual progress on-site during construction and compare it with the planned progress. These visual models help users in comparing ongoing site activities and planned activities. By simulating construction sequences and material flow in advance, potential issues related to progress and material control can be identified and resolved proactively, before construction even begins.

Accurate tracking and analysis of the actual (as-built) progress status of a construction project are essential for effective decision-making throughout the construction stage. Advanced BIM-enabled methods now automate the monitoring process by integrating daily photographic data and point cloud technologies. For instance, techniques such as structure-from-motion, multi-view stereo, and voxel-based scene labeling allow for the photorealistic reconstruction of dense 4D point cloud models. These models represent the actual site conditions and are aligned with industry foundation classes-based BIM models through a registration process, enabling seamless comparison between planned and actual progress (Golparvar-Fard et al. 2015).

Furthermore, machine learning algorithms, particularly those based on Bayesian probabilistic models, are used to automatically detect physical progress, even when visibility is partially obstructed. This enables real-time, activity-level tracking of construction status, making physical progress monitoring more precise and significantly less reliant on manual observation. The fusion of 3D remote sensing data with 4D BIM models supports an automated and accurate measurement of physical progress, ensuring that deviations from the planned are detected early and addressed promptly (Kim et al.2013).

Additionally, photogrammetric surveys have proven to be a valuable tool for progress assessment. These surveys generate high-resolution point clouds by combining disparity maps using semi-global matching algorithms. The resulting data is then compared against the 4D BIM model, allowing for a detailed analysis of construction deviations. This comparison involves calculating distances between point cloud elements and corresponding BIM components using a spatial grid system. Each grid cell is analyzed for coverage and assigned a confidence value, which indicates the likelihood that a component has been installed as planned. This method provides a reliable, automated basis for validating construction progress and supports early intervention when discrepancies are found (Braun et al. 2015). BIM-based progress monitoring, especially when combined with emerging technologies such as photogrammetry, computer vision, and machine learning, is transforming how construction sites are managed. These digital methods offer a comprehensive data-driven approach to ensure that projects remain on track, both in terms of schedule and quality.

6.4.2 *Tracking Construction Resources*

On large-scale infrastructure projects, where construction sites span vast areas and involve complex logistics, maintaining strict adherence to building protocols is essential. Effective materials tracking plays a pivotal role in ensuring operational efficiency, safety, and regulatory compliance. Real-time tracking systems that monitor the movement and status of materials, equipment, and personnel are critical for project managers to uphold standards related to safety, quality control, workforce management, and local ordinance compliance. One of the most effective methods for achieving this is the integration of radio frequency identification technology with BIM. This combination enables real-time identification and location tracking of construction resources across the job site. Radio frequency identification tags attached to equipment, tools, and even personnel provide continuous data streams that, when connected to the BIM environment, create a dynamic system for visualizing site activities and resource allocation. The real-time data generated from radio frequency identification-BIM integration supports the development of leading indicators for project management, such as safety performance metrics, security monitoring, and maintenance verification. This system not only enables proactive risk management but also supports real-time visualization of compliance with zoning rules, safety zone restrictions, and access control regulations. By providing a centralized and digital platform for tracking and analyzing movement on-site, radio frequency identification-BIM systems help enforce building protocols, reduce the potential for safety violations, and improve overall site management (Costin et al. 2015).

In addition to on-site tracking, modern construction projects, especially those utilizing off-site manufacturing approaches such as prefabricated steel, concrete panels, or modular units, demand efficient monitoring of materials across the entire supply chain. These components are fabricated in factories, transported to the job site, and installed according to the execution. The logistics involved in this process are complex, encompassing raw material procurement, fabrication timelines, shipping coordination, and on-site assembly. Any disruption can lead to delays, cost overruns, or material shortages. To address these challenges, BIM-enabled digital platforms are used to monitor construction materials from production to installation. These platforms track material flow in real-time and serve as collaborative tools that connect manufacturers, logistics providers, site managers, and financial stakeholders. By offering continuous updates on material status, delivery schedules, and installation progress, the system enhances transparency, supports just-in-time delivery, and reduces the risk of material mismanagement or wastage (Lwin 2024). Moreover, a real-time material tracking system supports informed decision-making by offering analytics and alerts that highlight bottlenecks, inventory imbalances, or scheduling conflicts. It strengthens coordination between supply chain partners and ensures alignment between the project's operational timeline and financial planning. As a result, BIM-integrated material tracking contributes significantly to project efficiency, cost control, and quality assurance, particularly in complex and multi-stakeholder construction environments.

6.4.3 *Quality Control and Record-Keeping*

Maintaining high-quality standards is critical in the construction industry, yet current practices used for construction quality management are often plagued by inefficiencies and complexities. Traditional quality management workflows are frequently hampered by time-consuming inspections, disorganized or incomplete documentation, poorly handled nonconformance reports, and fragmented communication among project stakeholders. These challenges are further intensified by inconsistencies in documentation procedures, missed quality checkpoints, and the cumbersome shift from paper-based records to digital formats. Together, these issues increase the risk of construction defects, compromise structural integrity, and heighten the potential for project delays and safety hazards. A major contributing factor to these inefficiencies is the absence of a standardized, transparent, and streamlined approach to managing quality-related information. Without a centralized system, the manual processes involved in recording, retrieving, analyzing, and sharing inspection data are not only resource-intensive and error-prone but also susceptible to miscommunication and even deliberate data manipulation. This lack of systemization ultimately undermines quality assurance efforts, reduces accountability, and increases the likelihood of disputes among contractors, consultants, and clients.

To address these challenges, a BIM-based quality management system presents a transformative solution. By embedding quality management functions directly into the BIM environment, a digital construction quality document management system becomes centralized, standardized, and highly transparent. Such a system ensures that all quality-related documents, including inspection reports, compliance checklists, nonconformance reports, and approval workflows, are securely stored and easily accessible within a unified digital framework. BIM also facilitates consistent documentation practices and strengthens data integrity, ensuring that quality assurance measures are applied uniformly across the project. Moreover, by automating data entry, audit trails, and reporting functions, the system reduces administrative overhead and minimizes human error. Ultimately, a BIM-enabled construction quality document management systems contribute to improved construction engineering and project management efficiency by promoting accountability, reducing rework, and enhancing overall project quality. It supports a proactive approach to quality control, where potential issues are identified and resolved earlier in the process, thereby reducing risk and improving stakeholder confidence (Ghosh and Karmakar 2025).

The large-scale data processing in road construction projects has the potential to significantly enhance the effectiveness and efficiency of quality control and quality assurance practices. By leveraging BIM, road construction monitoring can transition from traditional and reactive methods to a more proactive and data-driven approach. A BIM-based quality control and quality assurance system integrates real-time construction data directly into the information model, enabling seamless interaction between the digital model and underlying databases. This integration allows for the continuous collection, structuring, analysis, and visualization of construction data, which facilitates timely and informed decision-making throughout the construction stage (Han et al. 2023). At the core of this approach is the use of structured data formats and real-time data transmission methods, which ensure that all relevant information, from material usage and process compliance to environmental conditions, are captured and synchronized within the BIM environment. This centralized and dynamically updated model provides a reliable foundation for assessing construction quality against design and regulatory standards. Furthermore, advanced analytical models, including those based on the analytic hierarchy process, artificial neural networks, and genetic algorithms, can be embedded within the BIM framework. These models support intelligent quality control and quality assurance by identifying patterns, predicting deviations, and recommending corrective actions, thereby improving the responsiveness and accuracy of quality control processes across the construction stage.

BIM maintains a consistent linkage between design specifications, construction workflows, and quality benchmarks, enabling a coherent representation of how the physical construction process aligns with expected outcomes. This consistency is particularly advantageous for quality management, as it ensures that design data, execution schedule, and inspection records are all interconnected within the same model. 4D-BIM incorporates construction codes and standards, facilitating compliance verification in real-time. These models are often structured using a product-organization-process framework, which allows stakeholders to visualize the interdependencies among physical components, responsible parties, and construction sequences. This structured approach improves traceability, enhances accountability, and strengthens the alignment between the construction and quality management processes (Chen and Luo 2014). BIM transforms traditional quality management from a paper-based to a digital and real-time-based that supports continuous improvement and reduces the risk of defects and delays.

6.4.4 *Maintaining Health and Safety of Workers*

BIM technologies are increasingly being leveraged to enhance health and safety management within the construction industry. Using BIM for health and safety planning helps organizations in developing more sophisticated and effective health and safety management systems. This not only streamlines health and safety procedures but also boosts overall productivity, thereby strengthening the connection between health and safe practices and efficient project execution (Tender et al. 2022). BIM contributes significantly to cultivating a proactive safety culture during construction. This culture is essential for elevating the safety maturity levels of construction workers across various sites. A comprehensive safety culture encompasses three key components: a safety management system, individual and collective safety behavior, and the prevailing safety climate within the organization. The implementation of BIM enhances each of these dimensions by enabling better visualization, risk assessment, and communication, ultimately promoting a safer working environment. As a result, BIM-driven approaches not only improve immediate safety outcomes but also foster long-term behavioral and attitudinal shifts toward safety among construction personnel (Olugboyega and Windapo 2019).

6.4.5 *Delay Claim Analysis*

Delays and disruptions are frequent and often unavoidable in construction projects, particularly those involving complex scopes, multiple stakeholders, and tight schedules. As project complexity increases, so too does the difficulty of conducting accurate forensic schedule analysis and evaluating the extent and impact of delays. Factors such as concurrent delays, cascading secondary effects, and the cumulative impact of multiple delay and disruption events further complicate the process of determining accountability and justifying claims for time extensions or financial compensation. Establishing entitlement for delay-related claims requires a comprehensive demonstration of three key elements: causation (proving what specifically caused the delay), liability (identifying who is responsible), and quantum (quantifying the impact in terms of time and cost). Traditional methods of delay analysis, often reliant on static Gantt charts and fragmented documentation, are limited in their ability to accurately depict these dynamic and interrelated factors.

4D BIM has significantly transformed how delay and disruption are analyzed and visualized. By linking the project's 3D model with the construction schedule, 4D BIM allows stakeholders to simulate and visualize construction activities over time, including how and when delays occurred. This time-linked spatial visualization is especially useful in forensic schedule analysis, as it provides a clear, data-rich representation of the project's actual progress in comparison to the planned schedule. 4D BIM models help illustrate the sequence of work, highlight areas where activities were delayed or re-sequenced, and show how these delays impacted subsequent tasks. This allows claims analysts, legal professionals, and project managers to more accurately assess the ripple effects of delay events on the broader construction process. Furthermore, the simulation generates reliable data that can be used to support claims with greater objectivity and transparency (Valavanoglou and Heck 2016). By enhancing the accuracy, clarity, and credibility of delay and disruption analysis, BIM enables a shift from conventional, often subjective methods to a more evidence-based and visually intuitive approach. This not only improves the process of dispute resolution but also strengthens project planning and risk mitigation strategies in future construction efforts.

6.5 BIM and GIS Integration in the Construction Stage

The integration of BIM and GIS is used in many areas during the construction stage. The integration keeps track of the construction supply chain status and provides warning signals to ensure the delivery of construction materials on time (Irizarry et al. 2013). BIM provides detailed quantity takeoffs in an early phase of the procurement process and GIS supports a wide range of spatial analyses used in the logistics perspective of construction supply chain management. The application areas of BIM and GIS integration suggested in the literature include optimization of tower crane locations on large construction sites (Irizarry and Karan 2012), manifesting the flow of materials, availability of resources, and visual display of construction supply chain status (Deng et al. 2019). Other areas of BIM and GIS applications are low-disturbance construction (Sebastian et al. 2013), ensuring safety during construction (Zou et al. 2017), knowledge process of buildings (Vacca and Quaquero 2020), and documentation of infrastructure construction projects (Abd et al. 2020).

6.5.1 Site Monitoring

Despite technological and digital progress, project failure remains a significant and recurring challenge in the AEC industry, making it one of the most demanding and risk-prone sectors. Common challenges that lead to project failure include not ensuring worker safety, not meeting strict quality requirements, not addressing environmental preservation challenges, not complying with sustainability goals, and not managing privacy and legal regulations. These challenges often create delays, cost overruns, and efficiency losses. The introduction of artificial intelligence-powered drones offers new possibilities for overcoming many of these challenges. Artificial intelligence enables drones to recognize changes, particular patterns, and make informed assessments in real-time, transforming them from simple monitoring tools into decision-support systems. This innovation enhances risk detection, predicts maintenance, automates progress reporting, reduces human error, improves efficiency, and safety. The adoption of artificial intelligence-driven drones in the AEC industry is not just about achieving project success in terms of timelines and budgets. It also plays a vital role in safeguarding public health, protecting the environment, ensuring high-quality construction standards, and promoting economic sustainability. By reducing risks, minimizing waste, and ensuring compliance with sustainability requirements, artificial intelligence-powered drones pave the way toward a smarter, safer, and more sustainable built environment (Waqar et al. 2023).

Drones have emerged as powerful tools in the AEC industry, capable of capturing site-specific data with remarkable accuracy. The site-specific data significantly enhances BIM and GIS workflows by providing improved visualization, site analysis, and more efficient planning and management across all lifecycle stages of infrastructure projects. By offering aerial perspectives and real-time monitoring, drones facilitate a wide range of tasks, including 3D modeling for site assessment, site and infrastructure inspections, construction progress tracking, and real-time safety monitoring of workers and equipment on construction sites. Such applications improve decision-making, reduce operational costs, and increase precision by ensuring that all project stakeholders, planners, designers, contractors, and managers, have access to up-to-date and reliable data. A key advancement in this field is the integration of drone workflows with augmented reality and BIM.

An augmented reality-based solution for building inspection allows linking aerial drone footage with BIM data. This integration makes it easier to retrieve critical information, visualize risks, and better understand safety or structural concerns identified from drone-captured video footage. By combining augmented reality, drone footage, and BIM, project teams can more effectively identify potential hazards and take proactive measures to mitigate them (Liu et al. 2021).

6.6 Conclusions

The micro-level planning during the construction stage is a viable area to improve productivity. The use of GIS during the construction stage has an important impact on the project time duration, costs, and quality. GIS during the construction stage supports various types of geospatial analyses. The application areas of GIS during the construction stage start with the screening of a contractor based on his competency in doing work at a particular geographical location. GIS allows the integration of diverse datasets together to find a clear picture of the geographical trend of the labor market. Once construction starts, GIS is used to develop site layouts for materials storage and TFs. GIS use is also reported to locate heavy equipment on large construction sites by avoiding conflicts among them. In large infrastructure construction projects like the construction of highways, railways, canals, etc., where more earthwork is involved, GIS helps in planning for the volumes of soil that is cut and is dumped on the locations identified for filling. GIS is also used to develop an optimum earthmoving plan. Other application areas of GIS are site drainage planning, tracking site dynamics, and reducing wastage during construction.

BIM provides a centralized platform for creating, sharing, and integrating accurate project data, enabling real-time decisions that improve efficiency, safety, and sustainability. Its multi-dimensional modeling supports quality control, safety management, and carbon emission tracking. In underground and urban infrastructure construction, BIM links 3D/4D models with geotechnical data for excavation monitoring, risk management, and safety oversight. It also advances construction progress tracking, resource management, and as-built documentation. Additionally, BIM fosters a proactive health and safety culture. Further, BIM-GIS integration enhances supply chain management by tracking materials, issuing warnings, and supporting logistics with spatial analysis. Other applications include crane location optimization, materials flow visualization, safety assurance, low-disturbance construction, knowledge processes, and project documentation, improving efficiency, coordination, and decision-making.

6.7 Exercises

Question 6.1: Why is productivity during the construction stage considered important in the AEC industry?

Question 6.2: How do macro-level and micro-level planning contribute to productivity improvements during the construction stage?

Question 6.3: How does GIS assist in screening of contractors, and why are the geographical aspects important in contractor selection?

Question 6.4: What role does GIS play in human workforce planning, particularly in addressing location-specific labor shortages?

Question 6.5: How is GIS applied in construction materials trading, and what advantages does a web-based electronic market in GIS offer over other trading types?

Question 6.6: What safety considerations are addressed through GIS-based site layout planning to minimize accidents on construction sites?

Question 6.7: How does GIS contribute to earthwork planning, particularly in cut-and-fill volume calculations and optimizing haul routes?

Question 6.8: Describe how GIS supports progress tracking and site dynamics monitoring during the construction stage.

Question 6.9: How do GIS and GPS technologies together help in reducing construction waste at sites?

Question 6.10: In what ways does the integration of BIM with artificial intelligence help during the construction stage?

Question 6.11: How is BIM used in underground transportation infrastructure projects such as tunnels?

Question 6.12: How does BIM-based monitoring enhance risk management and safety in deep excavations?

Question 6.13: What role does BIM play in construction progress monitoring, and how do technologies like photogrammetry and machine learning contribute?

Question 6.14: How does the integration of BIM with RFID technology improve construction resources tracking and construction site management?

Question 6.15: What are the major applications of integrating BIM and GIS during the construction stage?

References

Abd, A. M., Hameed, A. H., and Nsaif, B. M. (2020). "Documentation of construction project using integration of BIM and GIS technique." *Asian Journal of Civil Engineering*, 21(7), 1249–1257.

Andoh, A., Su, X., and Cai, H. (2012). "A framework of RFID and GPS for tracking construction site dynamics." Proceeding of ASCE Conference on *Construction Research Congress-2012*, 818–827.

Anumba, C., Dainty, A., Ison, S., and Sergeant, A. (2008), "The utilisation of GIS in the construction labour market planning process." http://www.rics.org/NR/rdonlyres/B6AD4EB6-C418-4363-9DED-0997553D31C5/0/utilisation_of_gis.pdf.

Bansal, V. K. (2014). "Use of geographic information systems in spatial planning: A case study of an institute campus." *Journal of Computing in Civil Engineering*, 28(4), 1–12.

Bansal, V. K. (2017). "Integrated CAD and GIS-based framework to support construction planning: A case study." *Journal of Architecture Engineering*, 23(3).

Bansal, V. K. (2023). "A road-based 3D navigation system in GIS, a case study of an institute campus." *International Journal of Applied Geospatial Research (IJAGR)*, 14(1), 1–20.

Braun, A., Tuttas, S., Borrmann, A., and Stilla, U. (2015). "A concept for automated construction progress monitoring using BIM-based geometric constraints and photogrammetric point clouds." *Journal of Information Technology in Construction*, 20(5), 68–79.

Chen, A. Y., Mora, F. P., and Ouyang, Y. (2011). "A collaborative GIS framework to support equipment distribution for civil engineering disaster response operations." *Automation in Construction*, 20(5), 637–648.

Chen, L., and Luo, H. (2014). "A BIM-based construction quality management model and its applications." *Automation in Construction*, 46, 64–73.

Cheng, M. Y., and Chen, J. C. (2002). "Integrating barcode and GIS for monitoring construction progress." *Automation in Construction*, 11(1), 23–33.

Cheng, M. Y., and O'Connor, J. T. (1994). "Site layout of construction temporary facility using enhanced geographic information system (GIS)." *Automation in Construction*, 3(1), 11–19.

Cheng, M. Y., and O'Connor, J. T. (1996). "ArcSite: Enhanced GIS for construction site layout." *Journal of Construction Engineering and Management*, 122(4), 329–336.

Cheng, M. Y., and Yang, C. Y. (2001). "GIS-based cost estimate integrated with material layout planning." *Journal of Construction Engineering and Management*, 127(4), 291–299.

Costin, A. M., Teizer, J., and Schoner, B. (2015). "RFID and BIM-enabled worker location tracking to support real-time building protocol and data visualization." *Journal of Information Technology in Construction (ITcon)*, 20(29), 495–517.

Deng, Y., Gan, V. J. L., Das, M., Cheng, J. C. P., and Anumba, C. (2019). "Integrating 4D BIM and GIS for construction supply chain management." *Journal of Construction Engineering and Management*, 145(4), 04019016.

Ding, L., Zhou, Y., and Akinci, B. (2014). "Building information modeling (BIM) application framework: The process of expanding from 3D to computable nD." *Automation in Construction*, 46, 82–93.

Fabozzi, S., Biancardo, S. A., Veropalumbo, R., and Bilotta, E. (2021). "I-BIM based approach for geotechnical and numerical modelling of a conventional tunnel excavation." *Tunnelling and Underground Space Technology*, 108, 103723.

Ghosh, B., and Karmakar, S. (2025). "Development and application of a digitalized construction quality document management system for advanced construction using building information modeling." *Journal of Construction Engineering and Management*, 151(6), 04025053.

Golparvar-Fard, M., Pena-Mora, F., and Savarese, S. (2015). "Automated progress monitoring using unordered daily construction photographs and IFC-based building information models." *Journal of Computing in Civil Engineering*, 29(1), 04014025.

Han, C., Han, T., Ma, T., Tong, Z., and Wang, S. (2023). "A BIM-based framework for road construction quality control and quality assurance." *International Journal of Pavement Engineering*, 24(1), 2209903.

Hassanein, A. (2002). *Planning and scheduling highway construction using GIS and dynamic programming*. PhD Thesis, Concordia University, Canada.

Hegazy, T. (2005). "GIS support for optimizing infrastructure programs of distributed sites." Proceeding 10th International *Conference on Civil, Structural and Environmental Engineering Computing*, Civil-Comp Press, Stirling, Scotland.

Hegazy, T., and Elbeltagi, E. (1999). "EvoSite: Evolution-based model for site layout planning." *Journal of Computing in Civil Engineering*, 13(3), 198–206.

Irizarry, J., and Karan, E. P. (2012). "Optimizing location of tower cranes on construction sites through GIS and BIM integration." *Journal of Information Technology in Construction*, 17, 351–366.

Irizarry, J., Karan, E. P., and Jalaei, F. (2013). "Integrating BIM and GIS to improve the visual monitoring of construction supply chain management." *Automation in Construction*, 3, 1241–1254.

Jeljeli, M. N., Russell, J. S., Meyer, H. W. G., and Vonderohe, A. P. (1993). "Potential applications of geographic information systems to construction industry." *Journal of Construction Engineering and Management*, 119(1), 72–86.

Kang, S., and Seo, J. (2013). "GIS method for haul road layout planning in large earthmoving projects: Framework and analysis." *Journal of Construction Engineering and Management*, 139(2), 236–246.
Kim, C., Son, H., and Kim, C. (2013). "Automated construction progress measurement using a 4D building information model and 3D data." *Automation in Construction*, 31, 75–82.
Kumar, S., and Bansal, V. K. (2016). "A GIS-based methodology for safe site selection of a building in a hilly region." *Frontiers of Architectural Research*, 5(1), 39–51.
Kumar, S., and Bansal, V. K. (2019). "Use of GIS in locating TFs safely on a construction site in hilly regions." *International Journal of Construction Management*, 19(4), 341–353.
Kumar, S., and Bansal, V. K. (2021). "GIS-based locational evaluation of infrastructural facilities in hilly regions: A case study of an institute campus." *International Journal of Construction Management*, 21(11), 1165–1184.
Li, H., Chen, Z., Yong, L., and Kong, S. C. W. (2005). "Application of integrated GPS and GIS technology for reducing construction waste and improving construction efficiency." *Automation in Construction*, 14(3), 323–331.
Li, H., Kong, C. W., Pang, Y. C., Shi, W. Z., and Yu, L. (2003). "Internet-based geographic information systems system for e-commerce application in construction material procurement." *Journal of Construction Engineering and Management*, 129(6), 689–697.
Liu, D., Xia, X., Chen, J., and Li, S. (2021). "Integrating building information model and augmented reality for drone-based building inspection." *Journal of Computing in Civil Engineering*, 35(2), 04020073.
Lu, S. R., Wu, I. C., and Hsiung, B. C. B. (2012). "Applying building information modelling in environmental impact assessment for urban deep excavation projects." In *ISARC. Proceedings of the International Symposium on Automation and Robotics in Construction* (Vol. 29, p. 1). IAARC Publications.
Lwin, T. (2024). A BIM-enabled materials flow tracking platform for the construction supply chains of modular steel projects, Thesis of Master of Engineering, Chulalongkorn University,
Moselhi, O., and Alshibani, A. (2007). "Crew optimization in planning and control of earthmoving operations using spatial technologies." *ITcon*, 12, 121–137.
Moselhi, O., and Hassanien, A. (2003). "Data acquisition and analysis for highway construction using geographic information systems." *Canadian Journal of Civil Engineering*, 30, 533–542.
Nadeem, A., Wong, A. K. D., Akhanova, G., Azhar, S., and Wong, S. N. (2017). "Application of building information modeling (BIM) in site management—material and progress control." In *Proceedings of the 21st International Symposium on Advancement of Construction Management and Real Estate*, Singapore: Springer Singapore, 289–297.
Olugboyega, O., and Windapo, A. (2019). "Building information modeling—Enabled construction safety culture and maturity model: A grounded theory approach." *Frontiers in Built Environment*, 5, 35.
OpenAI (2025). ChatGPT (Version 5.1) [Large language model]. https://chat.openai.com/
Poku, S. E., and Arditi, D. (2006). "Construction scheduling and process control using geographic information systems." *Journal of Computing in Civil Engineering*, 20(5), 351–360.
Sebastian, R., Böhms, M., and van den Helm, P. (2013). "BIM and GIS for low-disturbance construction." In: N. Dawood and M. Kassem (Eds.), *Proceeding of 13th International Conference on Construction Applications of Virtual Reality*, 30-31 October 2013, London, UK, 469–479.
Seo, J., and Kang, S. (2006). "Geographic information system based roadway construction planning." *Canadian Journal of Civil Engineering*, 33(5), 508–520.

Su, X., Andoh, A. R., Cai, H., Pan, J., Kandil, A., and Said, H. M. (2012). "GIS-based dynamic construction site material layout evaluation for building renovation projects." *Automation in Construction*, 27, 40–49.
Tender, M., Couto, J. P., and Fuller, P. (2022). "Improving occupational health and safety data integration using building information modelling: An initial literature review." *Occupational and Environmental Safety and Health*, III, 75–84.
Vacca, G., and Quaquero, E. (2020). "BIM-3D GIS: An integrated system for the knowledge process of the buildings." *Journal of Safety Science*, 65(2), 193–208.
Valavanoglou, A., and Heck, D. (2016). "Building information modelling and forensic analysis of delay and disruption." In Komurlu, R., Gurgun, A.P., Singh, A. and Yazdani, S. (Eds), The First European and Mediterranean Structural Engineering and Construction Conference (EURO-MED-SEC-1), Istanbul, Turkey, ISEC Press.
Waqar, A., Othman, I., Hamah Sor, N., Alshehri, A. M., Almujibah, H. R., Alotaibi, B. S., and Aljarbou, M. (2023). "Modeling relation among implementing AI-based drones and sustainable construction project success." *Frontiers in Built Environment*, 9, 1208807.
Wu, I. C., Lu, S. R., and Hsiung, B. C. (2015). "A BIM-based monitoring system for urban deep excavation projects." *Visualization in Engineering*, 3(1), 11.
Zou, Y., Kiviniemi, A., and Jones, S. W. (2017). "A review of risk management through BIM and BIM-related technologies." *Safety Science*, 97, 88–98.

Chapter 7

The Operation and Maintenance Stage

7.1 Learning Objectives

After the completion of this chapter, readers will be able to:

- use GIS in infrastructure maintenance decision-making,
- explore the use of GIS in energy performance checking,
- explore the use of integration of BIM and GIS in the operation and maintenance stage, and
- understand the application areas of BIM and the integration of BIM and GIS in the O&M stage.

7.2 Introduction

The 3D model of an infrastructure along with its virtual surroundings in geographic information system (GIS) provides a virtual environment to organize various site operations during the operation and maintenance (O&M) stage. It provides visual information used in operation management, developing maintenance plans, and developing new/changing components of an infrastructure. For example, GIS is utilized for road infrastructure maintenance management works in which volumes or surface areas are calculated from virtual 3D model to estimate quantities of work involved at a particular location. The site constraints are easily modeled in GIS which play a major role in the smooth operation and maintenance of an infrastructure. The GIS-based scheduling model is used for optimizing the infrastructure maintenance program on distributed sites (Hegazy 2005). GIS use is explored for the selection of particular equipment under specific work conditions on different sites. The possibility of receiving information promptly enables project managers to proactively identify and resolve performance discrepancies before they incur additional costs and delays during disasters. GIS-based decision-support systems are developed and used for maximizing the water savings from efficiency improvements and deciding conservation measures through integrated management of large-scale irrigation systems (Triana and Labadie 2012).

DOI: 10.1201/9781003759348-7

GIS use is explored to investigate the wall effect caused by the proliferation of high-rise buildings in which the concept of building frontal area index, calculated based on existing buildings, is used (Wong et al. 2011). The ventilation pathways across the urban areas are located using raster-based least-cost path analysis in GIS. The winds are forced by high frontal areas to deviate around coastlines with blocks of wall effect causing buildings parallel to the coast. The overlay of ventilation pathways over a thermal satellite image, representing heat island intensity, indicates a significant reduction in heat island intensity values around the ventilation paths. GIS is also used as a disaster decision-support system based on real-time observations during infrastructure rehabilitation (Zerger and Smith 2003). This chapter, in the following sections, covers the application areas of GIS, building information modeling (BIM), and integration of BIM and GIS in the O&M stage of infrastructure projects:

7.3 GIS in O&M Stage

GIS aids in infrastructure maintenance by providing accurate information as and when needed. GIS enhances the infrastructure O&M stage by improving data access, utility management, health monitoring, flood damage assessment, retrofitting, emergency evacuations, etc., as discussed below:

7.3.1 Site Drainage Planning

GIS is used to identify stream channels for site drainage planning in case of heavy rainfalls, cloud bursts, or flooding of an infrastructure and its surroundings. Figure 7.1 shows a raster of accumulated rainwater flow to each cell by accumulating the weights of all cells that contribute rainwater flow into each downslope cell of a site. The identified stream channels to drain the site are shown in various shades of black color. The darker black color represents more rainwater inflow through that channel. The stream channels map is also useful in controlling soil erosion around an infrastructure. The stream channels having high rainwater flow values, as shown by the darker black color, cause more scouring during heavy rainfalls. Keeping this in view, a site drainage plan is developed in GIS to drain a site during heavy rainfalls. GIS is also used to find changes in the drainage pattern of a site if any alteration is proposed in its immediate surroundings.

7.3.2 Maintenance Decisions

7.3.2.1 Road Infrastructure

The timely maintenance of highways is essential for uninterrupted traffic. The governments are struggling to maintain highways in good working condition due to budget limitations and a lack of decision-making tools that help in prioritizing

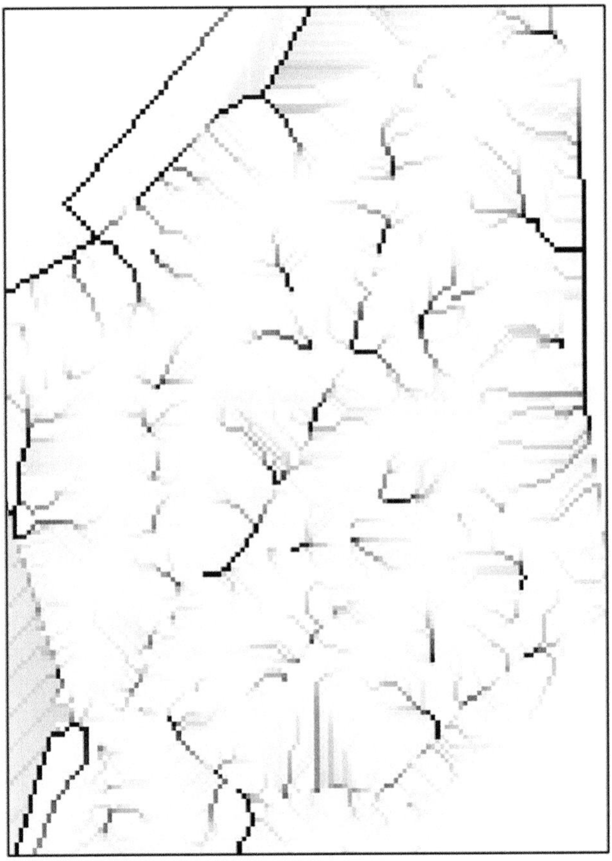

Figure 7.1 Identified stream channels to drain a site

highways that require immediate maintenance. GIS is used for the management of traffic, road pavement conditions, and safety data. GIS-based maintenance management assists in making reliable and accurate decisions regarding road maintenance and management, considering traffic counts, existing pavement conditions, maintenance costs, and alternative roadways available (Shrestha and Pradhananga 2009).

The deteriorating road conditions, increasing traffic loads, and decreasing funds have presented a complex management challenge for the road maintenance and rehabilitation process. There is a need to continuously assess the condition of the pavement network, decide on maintenance strategies, set rehabilitation priorities, and implement a maintenance management system. GIS is a powerful software tool where information on pavement sections stored in a textual database is linked to their locations. The GIS-based database is used to justify a budgetary request for the investment of public funds and help in infrastructure maintenance decisions (Mehdi et al. 2021).

GIS helps in the decision-making process for road infrastructure maintenance planning. Roadside slope failures often result in long traffic disruptions; hence, timely maintenance of roadside slopes needs greater priority for the smooth operation of traffic. GIS is used for the development of maintenance programs considering pavement condition and roadside slope stability. The roughness index is used as the pavement condition index. The roadside slope maintenance priority maps are produced by analyzing roadside slope stability. An integrated maintenance priority map is produced by combining both pavement condition and roadside slope maintenance priority maps with the consideration of the weighting of each maintenance component (Pantha et al. 2010). GIS also facilitates maintenance programming based on deterministic and probabilistic models to predict road pavement condition (Ribeiro et al. 2019). GIS is used in the development of the procedure for inspection and quality evaluation of road pavements. It is possible to establish and analyze the maintenance scenarios for road pavements.

Transportation asset management requires a cost-effective process to construct, operate, and maintain physical capital. It requires the allocation of resources, including personnel, equipment, materials, and supplies, during the O&M stage. GIS and artificial intelligence techniques are used together to develop an intelligent asset management system to optimize road and bridge maintenance programs. Artificial intelligence uses a human-like reasoning concept to optimize the transportation asset management system (Salim et al. 2002).

It is useful to develop visual maps of the road networks to monitor road conditions. The road profiler is generally used to collect the road profile data and the international roughness index is used for monitoring pavement roughness conditions. GIS is used for the development of the database of road maps and the visualization of the international roughness index data as an attribute. Different statistical characteristics of road profiles are easily depicted in GIS, which helps in developing a good pavement management system for the evaluation and management of different roads (Abulizi et al. 2016).

The road pavements deteriorate with time. The regular maintenance of pavements not only keeps them functional but also increases their service life. Hence, planning and scheduling of maintenance of pavements is an important part of the maintenance program. GIS is used for the development of maintenance plans and schedules for low-volume rural roads. The roads in the network are prioritized for pavement maintenance, depending on their pavement conditions and traffic volumes. A suitable repair and maintenance technique is selected for each road based on the predominant pavement defects, traffic volumes, and climatic conditions. The outcome of the maintenance plan is the priority ranking along with the most suitable repair technique for each road in a network (Nautiyal and Sharma 2021).

The maintenance of transportation infrastructure is continuously becoming a challenge for most of transportation agencies. GIS is used to manage maintenance operations for roads and parking lots, keeping in view the pavement condition index survey (Faris and Mahir 2012).

7.3.2.2 *Other Infrastructure*

Increasing demand for railway transportation requires large amounts of money to be spent on its maintenance and renewal. In industrialized and developing countries, one of the main problems of railway companies is regarding track maintenance and renewal. The cost reduction and better control of the maintenance process, therefore, increase the efficiency of the railway system. GIS is used to develop a decision-support system for analyzing the track components and suggesting methods for helping the track managers and engineers in deciding when maintenance is necessary and when is the best time for renewal. GIS enables users to visualize and analyze the railway track data, which is no longer possible using manual methods (Guler 2012).

Underground utility tunnels are crucial for urban areas with high density in terms of both population and number of buildings, since they tremendously improve the utilization of land resources. GIS is used in mine construction and maintenance planning to develop utility tunnels and minimize impacts on the environment. It pertains to mine developmental activities and the range of environmental activities to be managed. GIS is appropriate for spatial data management related to ore deposits and managing mine environment conditions (Choi et al. 2020).

There are many failures of river embankments following river floods. GIS is used for assessing the performance of embankments and to plan inspections, maintenance, and repair works. The failure mechanisms are modeled and performance indicators for each mechanism are identified. It is possible to model embankment structure, understand design, identify failure modes, and determine causes and effects. The performance of an embankment is assessed with respect to each mechanism by aggregating the corresponding indicators. It uses a rule-based multicriteria assignment method after discussions with a panel of experts. GIS is used to consider the vulnerability dimension, optimizing crisis management, and risk management (Serre et al. 2008). The lack of information regarding the condition of roads and irrigation facilities is one of the factors contributing to their poor maintenance program. A web-based GIS system as a part of the spatial decision-support system is suitable for planning maintenance of roads and irrigation facilities (Nainggolan et al. 2024).

7.3.3 *Inspection and Safety Planning*

The inspection of a pavement requires the involvement of manpower and expenditure; hence, the automatic inspection of pavement damage by image analysis is useful in order to reduce the burden of work involved. The crack ratio is one of the indices used to quantitatively evaluate the soundness of asphalt pavement. A convolutional neural network automatically detects and evaluates cracks from images of a pavement using deep learning. GIS represents the results graphically (Chun et al. 2021). GIS is used for road safety inspection during operation and finally to manage road infrastructure safety (Budzyński et al. 2018). There are many factors that influence accidents and their severity which are considered in ensuring road safety. The safety influencing factors are categorized into manmade, vehicular, and environmental. GIS helps in better understanding the road safety influencing factors, corresponding data acquisition, visualization of road safety influencing factors, and effective presentation of risk ranking. This is helpful in identifying high-risk road sections and supports effective communication in ensuring road safety.

7.3.4 *Infrastructure Resilience*

Infrastructure resilience is the ability of a structure to withstand and recover from the effects of natural or man-made hazards. Infrastructure resilience during and after any disruptive event is important for any community. The performance of an infrastructure varies from one location to another and the recovery process is also region-specific. A GIS Bayesian belief network is used to model and quantify the resilience in terms of vulnerability and recovery of road infrastructure against flood hazards (Kanti Sen and Dutta 2020). The data collection is done using a GIS-based platform and a probabilistic graphical model is used to model uncertainties in resilience quantification based on the available data and judgments.

7.3.5 *Energy Performance*

Assessing the hourly energy demand of existing building stock and predicting its variations due to energy efficiency measures is fundamental for planning strategies toward renewable energy. GIS is used in estimating the energy demand profiles of urban buildings based on volumetric building stock, their age of construction, and their widespread uses (Ferrari et al. 2021). GIS is used for energy performance mapping of heritage buildings, especially in the case of energy retrofit (Fabbri et al. 2012). GIS potential is used for the revaluation and measurement of town energy consumption which is related to the age and characteristics of existing buildings, both urban planning discipline and the heritage buildings discipline.

GIS is used to calculate solar irradiation on building surfaces by adopting commonly available base data on urban morphology, such as topographical maps or digital terrain models (Pili et al. 2018). The GIS-based approach is used for assessing thermal and electrical energy consumption and determining the usable solar energy potential of residential buildings in order to decrease the consumption of fossil fuels (Groppi et al. 2018).

7.3.6 *Environmental Impacts*

The GIS-based map overlay method is used to analyze the environmental vulnerability around roads and its impacts on the environment. GIS use is adapted for the comprehensive assessment of environmental impacts after the optimal selection of road alignment. It includes deciding the assessment factors and their weights, developing environmental vulnerability grade maps, calculating the coefficient of road impact extent for each assessment factor, and evaluating the alternative alignments comprehensively to obtain the best one (Li et al. 1999).

7.3.7 *Infrastructure Operations*

GIS is used to model a road network for network analysis which is quite useful during the operational stage. Generally, existing approaches related to impedance modeling in road network analysis include distance and time as the impedance variables. GIS is used to consider additional realistic impedance variables into account in the network analysis (Sadeghi-Niaraki et al. 2011). The realistic impedance variables include weather, sightseeing information, road types, and so on. The combined effect of these variables is determined in GIS using the analytical hierarchical process. It is observed that the route identified through the network analysis after incorporating the realistic impedance variables matches with the route people normally choose. Accident outcomes are usually the loss of lives, properties, and injuries on roads. GIS is used to analyze road accidents in different circumstances and road design that can be noteworthy in traffic accident prevention (Shahzad 2020). There is a need to improve the resilience of electrical distribution networks, looking at the reduction in frequency caused by extreme weather events. A large amount of data from different sources are collected and integrated to describe load profiles adequately. Machine learning and GIS are used for electrical load assessment to increase distribution network resilience (Bosisio et al. 2021).

7.4 BIM in O&M Stage

Effective decision-making in building maintenance requires the integration of diverse types of information and knowledge generated by members of the construction and facility management teams. This includes information and knowledge such as maintenance records, work orders, root causes of component failures, cascading effects of system failures, etc. When this information and associated knowledge are not systematically captured, stored, shared, or utilized, the result is often suboptimal decision-making, leading to increased operational costs, inefficiencies, and reduced building performance. Traditional building maintenance systems tend to emphasize either information capture or knowledge management, but rarely in an integrated fashion. Most current systems are employed during or after maintenance activities to document what has occurred, monitor building deterioration, and support subsequent corrective or preventive maintenance decisions. These systems rely on various information modeling techniques and knowledge management frameworks to enable comprehensive retrieval and application of maintenance-related information.

BIM serves as a foundational tool in this regard. BIM allows for the structured capture of relevant data (e.g., geometry, asset specifications, and component relationships), while knowledge-based systems enable reasoning about maintenance procedures, failure causes, and strategic planning. Integrated knowledge-based BIM platforms offer advanced functionalities that support construction and operational decisions through a seamless interplay of data and domain expertise (Motawa and Almarshad 2013). BIM presents a transformative solution for enhancing the O&M stage of the infrastructure lifecycle. It improves data accessibility for O&M personnel and enables the automated transfer of critical information into computerized maintenance management systems or computer-aided facility management systems. This automation facilitates the exchange of essential data such as asset location, spatial relationships, manufacturer specifications, and repair histories. BIM supports 3D visualization of components of infrastructure and understanding of their topological interrelations, critical for identifying and prioritizing maintenance tasks (Akcamete et al. 2010).

BIM enables the identification of operationally significant data, supports real-time decision-making, and manages uncertainty resulting from incomplete or outdated building documentation (McArthur 2015). It plays a pivotal role in maintenance management, which encompasses all services required to sustain the operational use of an infrastructure. BIM models are queried to extract repair and maintenance data and integrate data into relational databases and aligning with broader asset management systems (Akinradewo et al. 2023). For green buildings, which have more rigorous performance standards compared to conventional facilities, BIM supports higher-quality O&M processes. Its applications include energy management, security, emergency response, renovation planning, and space utilization. BIM thus contributes substantially to O&M of sustainable infrastructure (Cao et al. 2022).

BIM aids in damage detection and health monitoring of existing structures. Enhanced interoperability across BIM platforms, the adoption of advanced sensing technologies for fault detection, and the ability to manage large datasets, all while accounting for environmental impacts, improve the effectiveness of maintenance programs (Panah and Kioumarsi 2021). Structural health monitoring is particularly vital for public safety and economic stability. BIM-integrated health monitoring systems commonly apply offline data processing to analyze and visualize structural conditions. The Internet of Things (IoT) further enables real-time sensor data collection and integration. The emergence of BIM-enabled digital twins represents a major leap forward. These systems merge IoT sensing, wireless communication, digital signal processing, and structural engineering knowledge to offer dynamic monitoring and visualization of structural deformations. High-frequency updates in the digital twin environment allow for real-time analysis of how different components behave under varying load conditions (Hu et al. 2024).

BIM, when combined with IoT, autonomous robotics, and indoor positioning systems, enhances real-time indoor environmental monitoring. It offers high-resolution visualization of spatial data and environmental quality using standardized assessment indices. BIM digital twins accurately visualized air quality distributions in small to medium-sized spaces, while ultra-wideband positioning systems achieved centimeter-level accuracy for real-time tracking and assessment (Hu and Assaad 2024). Furthermore, the integration of BIM with artificial intelligence and machine learning unlocks new opportunities in predictive maintenance, cost and time estimation, energy modeling, and long-term infrastructure planning. These artificial intelligence-enhanced BIM systems improve the precision and reliability of construction-related estimations and performance forecasts (Heidari et al. 2024).

BIM serves as a central medium for effective communication and information exchange within the AEC industry during operation and maintenance. It enables accurate, timely, and context-appropriate sharing of information by directing the information flow through a centralized building model. BIM uses data-rich models that store extensive information, ranging from design specifications to real-time sensor data. This supports various analyses and simulations throughout the O&M stage of an infrastructure project. To interpret this information effectively for decision-making in the O&M stage, techniques from multiple visualization disciplines are applied.

Overall, BIM significantly contributes to the development of intelligent and automated maintenance systems. It supports retrofitting, infrastructure risk management, condition monitoring, and emergency planning simulations. The following sections further explore the applications of BIM during the O&M stage, illustrating how this technology continues to lifecycle management:

7.4.1 *Infrastructure Health Monitoring*

As infrastructure systems grow increasingly complex and expansive, the need for autonomous and intelligent condition assessment of large-scale structures, such as bridges, tunnels, and high-rise buildings, has become a critical focus in the field of civil and structural engineering. Over the past few decades, data-driven structural health monitoring techniques have gained prominence for their ability to deliver continuous and real-time insights into the structural integrity of such infrastructure. These techniques play a vital role in preserving structural safety, ensuring operational continuity, and enabling proactive risk and hazard mitigation through timely detection of problems and intervention in time. BIM has emerged as a powerful computational environment and a dynamic digital representation platform for integrating structural health monitoring systems. Serving as a centralized repository, BIM is capable of organizing, managing, and visualizing large volumes of sensor data and corresponding structural health indicators over time. This transforms traditional structural health monitoring workflows by providing spatial and semantic context to the sensor data, thereby facilitating more meaningful interpretations and informed decision-making.

BIM-enabled platforms are deployed to automate the ingestion and inventory of sensor data, such as strain, vibration, displacement, or temperature, directly into the BIM environment. This automation not only streamlines the process of condition assessment but also reduces the potential for human error commonly associated with manual inspection and data handling. By embedding sensor outputs into a BIM model, a system allows stakeholders to visually track the real-time condition of structural components and detect early signs of degradation or damage. Furthermore, the integration supports the development of systematic maintenance schedules and targeted risk management strategies, ensuring that remedial actions are timely and data-driven. In essence, the convergence of BIM and advanced structural health monitoring techniques offers a scalable, intelligent solution for the autonomous condition assessment of complex infrastructure. It enhances structural resilience, reduces maintenance costs, and supports the sustainable operation of critical assets under varying environmental and operational conditions (Boddupalli et al. 2019).

Seismic risk assessment is a critical requirement for many existing buildings, especially in cases involving changes in building usage, insurance evaluations, or re-occupancy authorization following a major seismic event. BIM plays an increasingly vital role in supporting the assessment and mitigation of earthquake-related risks by offering a centralized and data-rich environment that enhances both structural evaluation and emergency preparedness. BIM facilitates a comprehensive and reliable seismic assessment by providing detailed information on the physical and material properties of both structural components (like beams, columns, foundations, etc.) and non-structural components (like partitions, mechanical systems, etc.). This holistic data model enables engineers and risk analysts to perform more accurate simulations and vulnerability assessments, accounting for the complex interactions between various building systems during seismic events.

The integration of BIM with structural health monitoring technologies enables a self-diagnostic framework for buildings. This system processes real-time and post-event data on structural performance, such as strain, displacement, and damage indices, received from sensors deployed throughout the building. The use of BIM in this context significantly reduces reliance on traditional post-earthquake inspections, which are often labor-intensive, time-consuming, and potentially hazardous for inspectors. Instead, damage detection and evaluation are performed remotely, quickly, and with greater accuracy. Moreover, BIM supports the establishment of an emergency response management hub embedded within a building management system. This centralized hub facilitates real-time decision-making and control actions in the aftermath of an earthquake. For instance, it can be programmed to monitor the condition of critical mechanical and electrical systems and automatically initiate shutdown protocols for damaged or compromised services. This functionality is essential for minimizing the cascading effects of seismic damage, such as fire outbreaks, gas leaks, or equipment failures, thereby enhancing overall safety and resilience (Welch et al. 2014). Through these capabilities, BIM transforms seismic risk management from a reactive process to a proactive and automated system, improving both preparedness and response in earthquake-prone regions.

7.4.2 *Automated Maintenance*

Efficient maintenance strategies and informed decision-making approaches are often lacking in traditional facility management, leading to increased operational costs and reduced efficiency. To address this, BIM is increasingly adopted as a core tool in the development of automated facility maintenance management systems. BIM-based systems streamline the maintenance process by enabling the automatic scheduling of work orders, thereby improving decision-making and resource allocation. A crucial feature of BIM-based systems is their ability to allow bi-directional data exchange between BIM and facility maintenance management platforms. Through this integration, real-time information on maintenance work orders is extracted and visualized within the BIM environment via application programming interfaces. This visualization capability allows facility managers to identify and locate failed or deteriorating components within a building more effectively. BIM models store both geometric and semantic data of building components. When a component failure is detected, this data is extracted to compute an optimized maintenance plan. This is achieved using a modified version of Dijkstra's algorithm, which takes into account several parameters such as the type of issue, urgency level, spatial distance between affected elements, and their physical locations within an infrastructure (Chen et al. 2018). This algorithmic approach ensures that maintenance tasks are scheduled and executed in the most efficient manner possible.

The integration of BIM with digital twin technology has further enhanced real-time monitoring capabilities for urban infrastructure maintenance. This advanced system incorporates IoT sensor data, such as vibration, tilt, light intensity, air quality, and water presence, which are continuously synchronized, processed, and presented on a web-based dashboard. This setup enables the automated detection of anomalies, including unusual structural shifts or lighting conditions, thereby supporting timely and informed maintenance decisions with improved safety and operational reliability (Yang et al. 2025). Maintaining a comfortable indoor thermal environment is essential for occupant health, productivity, and overall well-being. Traditionally, building management systems are used to monitor and regulate the thermal conditions of indoor spaces. However, one of the main limitations of conventional building management systems lies in their reliance on non-interactive visualizations, such as static drawings, which hinder intuitive data interpretation.

A more dynamic and integrated solution is developed that combines sensor-based alert systems with BIM models. This approach enables real-time virtual visualization of a building's thermal conditions throughout its operational phase. The integration of BIM platforms with IoT devices allows for the continuous collection of environmental data, specifically temperature and humidity, which are transmitted to a centralized database server for analysis. A structured and automated workflow is then applied to compile, standardize, and integrate this data within the BIM environment. The resulting self-updating BIM model enables stakeholders to visualize and analyze environmental conditions in real-time. This facilitates efficient data interpretation and seamless exchange of monitoring information across systems. Ultimately, this provides a powerful and practical tool for intelligent environmental monitoring, enhancing decision-making capabilities for building operators, and supporting the maintenance of indoor thermal comfort at optimal levels (Valinejad-shoubi et al. 2021).

7.4.3 *Automated O&M of Smart Buildings*

The evolution of smart buildings stems from the progressive integration of advanced information and communication technologies with building systems and operations. This continuous digital enhancement enables buildings to be managed, monitored, and controlled remotely and intelligently throughout their lifecycle. Smart buildings aim to deliver optimized performance in terms of energy efficiency, occupant comfort, operational convenience, environmental sustainability, and cost-effectiveness. At the heart of this transformation lies the fusion of several cutting-edge technologies, including BIM, the IoT, machine learning algorithms, cloud computing, big data analytics, and sensor networks. Together, these technologies form the foundation for buildings that can make autonomous decisions and adapt to changing environmental and operational conditions in real-time.

BIM plays a pivotal role in the realization of smart buildings by providing detailed, data-rich 3D digital models that represent both the physical and functional aspects of a structure. These models not only offer geometric representations but also incorporate extensive project-related metadata, such as material specifications, equipment details, maintenance schedules, and spatial relationships. This holistic visualization and information-sharing capability enables all project stakeholders to gain a comprehensive understanding of the building, which in turn enhances collaboration, reduces errors, and improves project outcomes. The adoption of BIM has been shown to significantly reduce operational costs and elevate the overall quality and sustainability of built environments (Yang et al. 2021). As such, BIM is considered a fundamental component in the operation of smart buildings.

The concept of digital twins has emerged as a transformative layer in smart building development. Digital twins serve as dynamic and virtual counterparts of physical buildings, replicating both static attributes (geometry, structure, etc.) and dynamic behaviors (energy usage, occupancy patterns, etc.) through real-time data synchronization. This virtual-physical connectivity enables continuous monitoring of building performance, future scenarios, and proactive decision-making to optimize building operations. Despite its potential, one of the key challenges in implementing digital twin technologies within smart buildings is the lack of interoperability, particularly between BIM platforms and heterogeneous IoT data sources (Zhang et al. 2015). To address this limitation, research and development efforts have focused on multi-layer BIM-IoT data integration frameworks that ensure semantic interoperability. Such integration enables seamless communication and data exchange across various smart building systems, thereby supporting a unified digital twin environment capable of intelligent analysis, automation, and system-wide optimization (Eneyew et al. 2022). The smart buildings represent the convergence of architectural design, digital technologies, and intelligent systems. BIM, when integrated with IoT, digital twins, and data analytics, contributes to the creation of buildings that are not only efficient and resilient but also responsive and future-ready.

Moreover, BIM acts as a common language among all stakeholders and disciplines in a project, fostering collaboration and integrated project delivery. Its role in coordinating project systems mirrors that of a project manager. BIM integrates different disciplines through its robust information database, facilitates communication, supports constructability analysis, performs cost and time estimations via quantity takeoffs before and during the O&M stage, enables visualizations that provide a holistic view of a project, and helps build collaborative project teams (Rokooei 2015). Ultimately, BIM integrates critical components of infrastructure, such as risk, time, cost, energy, safety, and sustainability, into a unified platform (Ershadi et al. 2021).

7.4.4 *Planning for Retrofitting*

BIM serves as a highly effective platform for integrating sustainable design strategies into the renovation and retrofit of existing infrastructure. Its capability to support data-rich modeling and simulation makes it an ideal tool for embedding sustainability considerations into every phase of the retrofit process. BIM facilitates the analysis, decision-making, and implementation of environmentally responsible design concepts tailored specifically for the refurbishment of aging infrastructure (Hammond et al. 2014). Beyond facilitating design decisions, BIM is also instrumental in developing systematic evaluation frameworks and compiling sets of best practices for projects aiming to achieve sustainable performance standards. In the context of energy retrofitting, BIM supports the creation of comprehensive plans aimed at enhancing the energy efficiency of existing buildings, thereby contributing to broader sustainability goals.

A particularly impactful approach involves the integration of BIM with optimization algorithms, multi-dimensional modeling, and lifecycle assessment. This integrated methodology enables the evaluation of multiple retrofit scenarios, helping stakeholders identify energy-efficient alternatives based on lifecycle cost, performance, and environmental impact. As a result, decision-makers and investors are equipped with actionable insights that streamline the selection of effective retrofit strategies and sustainable materials. BIM also facilitates comparative analysis by modeling and simulating the thermal behavior of building components, specifically heating and cooling loads, both before and after the implementation of retrofit measures. This analysis allows a quantitative assessment of how retrofitting interventions, such as improvements to building envelopes or the installation of evaporative cooling systems, impact overall energy consumption. Ultimately, this data-driven, BIM-based approach yields optimal retrofit scenarios that reduce energy usage and environmental impact while improving building performance. It empowers stakeholders to make well-informed and sustainability-oriented decisions that align with long-term energy efficiency and environmental objectives (Motalebi et al. 2022).

7.4.5 *Planning for Emergency Situations*

In emergency scenarios such as fires, security threats, or natural disasters, effective response depends heavily on timely communication, accurate localization of first responders, and efficient coordination of tasks. These challenges are further amplified in complex and large-scale facilities where harsh conditions and low visibility hinder traditional navigation and communication methods. In such contexts, BIM offers powerful capabilities for supporting emergency planning, simulation, and response management. BIM includes detailed representations of physical structures, such as walls, doors, staircases, and systems, as well as virtual attributes like room functions, material properties, and access constraints. By providing a spatially accurate and information-rich model of a building's internal and external environments, BIM serves as an ideal foundation for developing emergency simulation tools and safety response strategies.

Recent advancements have enabled the creation of stand-alone BIM-based indoor localization systems, which facilitate the automatic extraction, transformation, and visualization of relevant building data for public safety applications (Feng et al. 2021). These systems map out room geometries, entry points, escape routes, and mechanical systems, allowing first responders to navigate efficiently and make informed decisions in real-time. Such localization and tracking capabilities are particularly critical in emergencies where seconds matter and visibility may be compromised. The comprehensive datasets stored in BIM models support the simulation of building operations and human behavior under both routine and emergency conditions. This includes evacuation modeling, fire and smoke propagation simulations, and dynamic crowd movement analysis. These predictive simulations are vital for identifying potential hazards, testing emergency protocols, optimizing escape routes, and allocating emergency resources effectively.

Beyond immediate response, BIM also contributes to post-event recovery and reconstruction efforts. By providing detailed as-built data, BIM expedites damage assessment, aids in structural analysis, and streamlines the rebuilding process. It ensures continuity and accuracy across planning, execution, and documentation phases, facilitating a more resilient and coordinated recovery (Drogemuller 2015). BIM functions as a critical platform for emergency preparedness, response coordination, and disaster recovery. Its ability to integrate spatial intelligence with real-time data and simulation tools makes it an invaluable asset for enhancing building safety and operational resilience.

7.5 BIM and GIS Integration in O&M Stage

7.5.1 *Developing Maintenance Programs*

The low accessibility of information regarding the design, construction, and current status of buildings is the main cause of inefficient action on existing buildings during the O&M stage. Further, the lack of as-built documentation, intended use, space dimensions, environmental context, maintenance history, compliance with current regulations, etc., causes difficulties in planning, scheduling, and controlling appropriate interventions during the O&M stage. The integration of BIM and GIS is used to structure the information used to support the operation and management of large building stocks (Vacca et al. 2018).

An infrastructure cannot be used for its intended purpose without the required utilities. The maintenance workflows require work to be performed inside and outside of an infrastructure. Therefore, it is necessary to get information about the inside of an infrastructure and outside utilities networks in advance to perform maintenance work efficiently. The information about the inside of an infrastructure is maintained and extracted from a building information model and outside information is extracted from GIS. GIS is a platform that integrates maintenance workflows for mechanical, electrical, and plumbing systems of an infrastructure (Liu and Issa 2012).

GIS facilitates decision-making for utility maintenance works. GIS is used to improve underground urban utility management efficiency from the perspective of utility components and utility networks. The main reasons for inefficient utility maintenance are poor information sharing, low utility management, and improper decision-making. BIM and GIS together represent utility information comprehensively to support underground utility management in terms of utility components and utility networks. The information-sharing process, utility management efficiency, and decision-making are improved and facilitated with the integration of BIM and GIS (Wang et al. 2019).

The utility tunnels are also operated via computer-based maintenance management systems in which integration of BIM and GIS is used to improve the performance of the maintenance management system. The integration helps in developing an effective maintenance program based on the updated information (Lee et al. 2018). The web-based bridge management system is developed through the integration of BIM and GIS (Zhu et al. 2021). The system is confined to a small number of areas, such as fire response management, traffic noise assessment, flood damage assessment, and indoor route planning.

The need for a comprehensive management and maintenance system to maintain underground utilities in a safe condition is becoming more prominent (OpenAI 2025). There are several problems with the current operation and maintenance practices in the underground utilities. The main problem is the lack of a digitalized 3D model with a comprehensive database for dynamically editing data on a site. A pipeline operation and maintenance management system by using BIM, GIS, augmented reality, integrating various project information from different sources through an application programming interface, and a cloud database was developed by Shekargoftar et al. (2022). It facilitates decision-making in the O&M stage through 3D visualizations, integration of construction information, maintaining inspection information, and developing efficient data retrieval mechanisms.

7.5.2 *Infrastructure Health Monitoring*

Infrastructure health has become an increasingly popular topic in recent years. The health of an infrastructure is directly related to the safety of its users. A scientific evaluation of the health of an infrastructure helps in assessing health issues and avoiding the catastrophic consequences of an infrastructure failure. The capabilities of the analytic hierarchy process, BIM, and GIS are used to develop an innovative and practical approach to evaluate building health (Ding et al. 2020). The building health evaluation model in the analytic hierarchy process considers five aspects: comfort performance, safety performance, environmental performance, operation management, and economic performance. Among them, safety performance is considered the most important factor in building a health evaluation. The energy consumption of a building is reduced by improving the economic performance and operational management performance.

7.5.3 *Conservation of Historical Infrastructure*

An architectural heritage structure carries heterogeneous and multi-layered information beyond its physical characteristics. Its conservation and O&M require an integrated representation of various types of information for understanding and management before the decision-making process. BIM is used widely for graphical information management of many historical buildings; however, information management of such buildings requires attribute management and integration of different spatial and non-spatial datasets. GIS has great potential in spatial and non-spatial dataset management and modeling of spatial relationships, but limited 3D representation and editing capabilities. Saygi et al. (2013) evaluated the role of BIM and GIS together, highlighting their advantages and disadvantages for integration, retrieval, and management of heterogeneous data in the context of historical buildings for their conservation and O&M.

Multidisciplinary data integration within an information system is considered a key point for infrastructure rehabilitation projects. Information regarding the state of preservation and/or decision-making is a prerequisite for sustainable restoration. In addition to achieving the structural integrity of a historic building, especially one that has undergone many construction phases, developing its restoration inventory is a very elaborate task that involves multidisciplinary information regarding history, architecture, materials used, and geometric data. BIM and GIS are used for data acquisition, classification, and management of various multisensory data for modeling and monitoring of historic buildings (Tsilimantou et al. 2020). The assessment is done qualitatively and quantitatively to facilitate the structural assessment of historic buildings.

7.5.4 *Damage Assessment*

Flood damage assessment is a key component of risk management. It provides the basis for decision-making and treatment of risks. The information required for micro-level flood damage assessment analysis includes building information represented in BIM and flood damage information represented in GIS. BIM and GIS cannot independently represent all the information required in flood damage assessment. The integration of BIM and GIS is used in flood damage assessment to building infrastructure (Amirebrahimi et al. 2016).

7.5.5 *Retrofitting*

In response to rising energy costs and motivation to reduce environmental impacts, upgrading the large building stock that is responsible for more energy consumption to minimize energy consumption is becoming an important task. Building retrofitting is done to make buildings more energy efficient. Göçer et al. (2016) introduced a pre-retrofit model that efficiently integrates multiple forms of building data to develop a comprehensive understanding of buildings to be renovated. BIM and GIS are used to represent complex information about buildings for retrofitting and conducting scenario-based comparisons and optimizing renovation decisions based on economic and environmental goals.

7.5.6 *Planning Emergency Evacuations*

The integration of BIM and GIS is useful for integrating the internal environment of an infrastructure and road networks for designing optimal routes in emergency evacuations during urban disasters. One main challenge in disaster management and emergency evacuations is to find optimum routes and guide people through optimal routes to reach at safe locations. GIS is used to develop an algorithm to find optimum routes between different locations (Atyabi et al. 2019). The criteria of flammability, occupancy load, width of exit access, and distance of fire point are considered to find the safest routes during emergency evacuations.

7.6 Conclusions

GIS allows the management of information related to an infrastructure, understanding information with respect to the location of an infrastructure, and representing constraints. The maintenance professionals inspect an infrastructure, collect information about it at a site, and use mobile GIS to maintain information that is used to keep an infrastructure in good and working condition. The information related to an infrastructure is mapped in GIS and narrowed down to the required level of detail to forecast the type of maintenance required and associated costs at a site by identifying components that need maintenance. The information database is developed and used for hazard identification, calculation of risks, and forecasting the future condition of an infrastructure. The integration of GIS with other leading infrastructure management software tools is also possible to strengthen operation and maintenance programs. The location-based operations in GIS are used to develop real-time management capabilities to integrate data from finance, users, sensors, and planning professionals for improved decision-making.

BIM enables structured data capture and its integration with maintenance and asset management systems. When combined with IoT, artificial intelligence, and digital twins, BIM enhances predictive maintenance, real-time monitoring, and structural health assessments, supporting energy efficiency, safety, and sustainability. BIM applications span retrofitting, seismic risk management, emergency response, and maintenance of smart buildings. BIM-GIS integration enhances O&M by improving maintenance workflows, utility management, and decision-making through 3D visualization, cloud databases, and digitalized models. Applications of BIM-GIS integration include infrastructure health evaluation, historical conservation, flood damage assessment, retrofitting, and emergency evacuations.

7.7 Exercises

Question 7.1: How does GIS provide a virtual environment for planning and organizing site operations during the O&M stage of infrastructure projects?

Question 7.2: Develop a project in GIS for road infrastructure maintenance management.

Question 7.3: How do GIS-based scheduling models optimize maintenance programs for distributed infrastructure sites?

Question 7.4: What role does GIS play in decision-support systems during infrastructure rehabilitation?

Question 7.5: How is GIS applied in drainage planning in the event of flooding?

Question 7.6: Give application areas of GIS in the maintenance and renewal of railways, tunnels, and river embankments.

Question 7.7: What is the significance of integrating BIM with IoT and digital twin technologies for infrastructure health monitoring and automated maintenance?

Question 7.8: How does BIM contribute to the development of smart buildings, and what technologies work alongside it to achieve autonomous and intelligent operations?

Question 7.9: Why is the interoperability between BIM platforms and heterogeneous IoT data sources considered a major challenge in implementing digital twin technologies?

Question 7.10: How does the integration of BIM and IoT contribute to creating future-ready and resilient smart buildings?

Question 7.11: How is BIM utilized in emergencies such as fires or natural disasters to support evacuation planning and response management?

Question 7.12: What advantages does the integration of BIM and GIS provide in developing effective maintenance programs during the O&M stage?

Question 7.13: How is integration of BIM and GIS applied in infrastructure health monitoring?

Question 7.14: What are the challenges in conserving historical infrastructure, and how do BIM and GIS together address these challenges?

Question 7.15: How is integration of BIM and GIS used in flood damage assessment and retrofitting to support risk management and sustainable renovation?

Question 7.16: In the context of emergency evacuations during disasters, how does the integration of BIM and GIS help in determining optimal and safe evacuation routes?

References

Abulizi, N., Kawamura, A., Tomiyama, K., and Fujita, S. (2016). "Measuring and evaluating of road roughness conditions with a compact road profiler and ArcGIS." *Journal of Traffic and Transportation Engineering (English Edition)*, 3(5), 398–411.

Akcamete, A., Akinci, B., and Garrett, J. H. (2010). "Potential utilization of building information models for planning maintenance activities." *Proceedings of the International Conference on Computing in Civil and Building Engineering*, 151–157.

Akinradewo, O., Aigbavboa, C., Oke, A., Edwards, D., and Kasongo, N. (2023). "Key requirements for effective implementation of building information modelling for maintenance management." *International Journal of Construction Management*, 23(11), 1902–1910.

Amirebrahimi, S., Rajabifard, A., Mendis, P., and Ngo, T. (2016). "A BIM-GIS integration method in support of the assessment and 3D visualization of flood damage to a building." *Journal of Spatial Science*, 61(2), 317–350.

Atyabi, S., Kiavarz Moghaddam, M., and Rajabifard, A. (2019). "Optimization of emergency evacuation in fire building by integrated BIM and GIS." *The International Archives of the Photogrammetry, Remote Sensing and Spatial Information Sciences*, 42, 131–139.

Boddupalli, C., Sadhu, A., Rezazadeh Azar, E., and Pattyson, S. (2019). "Improved visualization of infrastructure monitoring data using building information modeling." *Structure and Infrastructure Engineering*, 15(9), 1247–1263.

Bosisio, A., Moncecchi, M., Morotti, A., and Merlo, M. (2021). "Machine learning and GIS approach for electrical load assessment to increase distribution networks resilience." *Energies*, 14(14), 4133.

Budzyński, M., Kustra, W., Okraszewska, R., Jamroz, K., and Pyrchla, J. (2018). "The use of GIS tools for road infrastructure safety management." *E3S Web of Conferences*, 26(9), EDP sciences.

Cao, Y., Kamaruzzaman, S. N., and Aziz, N. M. (2022). "Building information modeling (BIM) capabilities in the operation and maintenance phase of green buildings: A systematic review." *Buildings*, 12(6), 830.

Chen, W., Chen, K., Cheng, J. C., Wang, Q., and Gan, V. J. (2018). "BIM-based framework for automatic scheduling of facility maintenance work orders." *Automation in Construction*, 91, 15–30.

Choi, Y., Baek, J., and Park, S. (2020). "Review of GIS-based applications for mining: Planning, operation, and environmental management." *Applied Sciences*, 10(7), 2266.
Chun, P. J., Yamane, T., and Tsuzuki, Y. (2021). "Automatic detection of cracks in asphalt pavement using deep learning to overcome weaknesses in images and GIS visualization." *Applied Sciences*, 11(3), 892.
Ding, Z., Niu, J., Liu, S., Wu, H., and Zuo, J. (2020). "An approach integrating geographic information system and building information modelling to assess the building health of commercial buildings." *Journal of Cleaner Production*, 257, 120532.
Drogemuller, R. (2015). "BIM support for disaster response." *Risk-Informed Disaster Management: Planning for Response, Recovery and Resilience*, 391–405.
Eneyew, D. D., Capretz, M. A., and Bitsuamlak, G. T. (2022). "Toward smart-building digital twins: BIM and IoT data integration." *IEEE Access*, 10, 130487–130506.
Ershadi, M., Jefferies, M., Davis, P., and Mojtahedi, M. (2021). "Implementation of building information modelling in infrastructure construction projects: A study of dimensions and strategies." *International Journal of Information Systems and Project Management*, 9(4), 43–59.
Fabbri, K., Zuppiroli, M., and Ambrogio, K. (2012). "Heritage buildings and energy performance: Mapping with GIS tools." *Energy and Buildings*, 48, 137–145.
Faris, F. G., and Mahir, M. D. (2012). "Using of modern GIS in road condition index." *Journal of Advanced Science and Engineering Research*, 2(3), 178–190.
Feng, Y., Wang, J., Fan, H., and Hu, Y. (2021). "A BIM-based coordination support system for emergency response." *IEEE Access*, 9, 68814–68825.
Ferrari, S., Zagarella, F., Caputo, P., and Dall'O', G. (2021). "A GIS-based procedure for estimating the energy demand profiles of buildings towards urban energy policies." *Energies*, 14(17), 5445.
Göçer, Ö, Hua, Y., and Göçer, K. (2016). "A BIM-GIS integrated pre-retrofit model for building data mapping." In *Building Simulation*. Springer Berlin Heidelberg, Germany.
Groppi, D., de Santoli, L., Cumo, F., and Garcia, D. A. (2018). "A GIS-based model to assess buildings energy consumption and usable solar energy potential in urban areas." *Sustainable Cities and Society*, 40, 546–558.
Guler, H. (2012). "Geographic information system-based railway maintenance and renewal system." *Proceedings of the Institution of Civil Engineers-Transport*, 165(4), 289–302.
Hammond, R., Nawari, N. O., and Walters, B. (2014). "BIM in sustainable design: Strategies for retrofitting/renovation." In *Computing in Civil and Building Engineering*, 1969–1977.
Hegazy, T. (2005). "GIS support for optimizing infrastructure programs of distributed sites." *Proceedings 10*th *Int. Conf. on Civil, Structural and Environmental Engineering Computing*, Civil-Comp Press, Stirling, Scotland.
Heidari, A., Peyvastehgar, Y., and Amanzadegan, M. (2024). "A systematic review of the BIM in construction: From smart building management to interoperability of BIM and AI." *Architectural Science Review*, 67(3), 237–254.
Hu, X., and Assaad, R. H. (2024). "A BIM-enabled digital twin framework for real-time indoor environment monitoring and visualization by integrating autonomous robotics, LiDAR-based 3D mobile mapping, IoT sensing, and indoor positioning technologies."*Journal of Building Engineering*, 86, 108901.
Hu, X., Olgun, G., and Assaad, R. H. (2024). "An intelligent BIM-enabled digital twin framework for real-time structural health monitoring using wireless IoT sensing, digital signal processing, and structural analysis." *Expert Systems with Applications*, *252*, 124204.
Kanti Sen, M., and Dutta, S. (2020). "An integrated GIS-BBN approach to quantify resilience of roadways network infrastructure system against flood hazard." *ASCE-ASME Journal of Risk and Uncertainty in Engineering Systems, Part A: Civil Engineering*, *6*(4), 04020045.

Lee, P. C., Wang, Y., Lo, T. P., and Long, D. (2018). "An integrated system framework of building information modelling and geographical information system for utility tunnel maintenance management." *Tunnelling and Underground Space Technology*, 79, 263–273.

Li, X., Wang, W., Li, F., and Deng, X. (1999). "GIS-based map overlay method for comprehensive assessment of road environmental impact." *Transportation Research Part D: Transport and Environment*, 4(3), 147–158.

Liu, R., and Issa, R. (2012). "3D visualization of sub-surface pipelines in connection with the building utilities: integrating GIS and BIM for facility management." *Proceeding of International Conference on Computing in Civil Engineering*, 341–348.

McArthur, J. J. (2015). "A building information management (BIM) framework and supporting case study for existing building operations, maintenance and sustainability." *Procedia Engineering*, 118, 1104–1111.

Mehdi, M. A., Cherradi, T., El Karkouri, S., and Qachar, A. (2021). "Applying Geographic Information Systems (GIS) for surface condition indicators modeling of a flexible pavement." *E3S Web of Conferences*, EDP Sciences, 298, 04001.

Motalebi, M., Rashidi, A., and Nasiri, M. M. (2022). "Optimization and BIM-based lifecycle assessment integration for energy efficiency retrofit of buildings." *Journal of Building Engineering*, 49, 104022.

Motawa, I., and Almarshad, A. (2013). "A knowledge-based BIM system for building maintenance." *Automation in construction*, 29, 173–182.

Nainggolan, T., Sai, S. S., Lomi, A., Mabrur, A. Y., and Andinisari, R. (2024). "Web-based GIS spatial decision support system for infrastructural maintenance of roads and irrigation facilities in Central and East Sumba." In *AIP Conference Proceedings*, 3077(1). AIP Publishing.

Nautiyal, A., and Sharma, S. (2021). "Condition-based maintenance planning of low volume rural roads using GIS." *Journal of Cleaner Production*, 312, 127649.

OpenAI (2025). ChatGPT (Version 5.1) [Large language model]. https://chat.openai.com/

Panah, R. S., and Kioumarsi, M. (2021). "Application of building information modelling (BIM) in the health monitoring and maintenance process: A systematic review." *Sensors*, 21(3), 837.

Pantha, B. R., Yatabe, R., and Bhandary, N. P. (2010). "GIS-based highway maintenance prioritization model: an integrated approach for highway maintenance in Nepal mountains." *Journal of Transport Geography*, 18(3), 426–433.

Pili, S., Desogus, G., and Melis, D. (2018). "A GIS tool for the calculation of solar irradiation on buildings at the urban scale, based on Italian standards." *Energy and Buildings*, 158, 629–646.

Ribeiro, A. M. G., Capitão, S. D., and Correia, R. G. (2019). "Deciding on maintenance of small municipal roads based on GIS simplified procedures." *Case Studies on Transport Policy*, 7(2), 330–337.

Rokooei, S. (2015). "Building information modeling in project management: necessities, challenges and outcomes." *Procedia-Social and Behavioral Sciences*, 210, 87–95.

Sadeghi-Niaraki, A., Varshosaz, M., Kim, K., and Jung, J. J. (2011). "Real world representation of a road network for route planning in GIS." *Expert Systems with Applications*, 38(10), 11999–12008.

Salim, M. D., Strauss, T., and Emch, M. (2002). "A GIS-integrated intelligent system for optimization of asset management for maintenance of roads and bridges." In *International Conference on Industrial, Engineering and Other Applications of Applied Intelligent Systems*, 628–637, Berlin, Heidelberg: Springer Berlin Heidelberg.

Saygi, G. A. M. Z. E., Agugiaro, G., Hamamcıoğlu-Turan, M., and Remondino, F. (2013). "Evaluation of GIS and BIM roles for the information management of historical buildings." *ISPRS Annals of the Photogrammetry, Remote Sensing and Spatial Information Sciences*, 2, 283–288.

Serre, D., Peyras, L., Tourment, R., and Diab, Y. (2008). "Levee performance assessment methods integrated in a GIS to support planning maintenance actions." *Journal of Infrastructure Systems*, 14(3), 201–213.

Shahzad, M. (2020). "Review of road accident analysis using GIS technique." *International Journal of Injury Control and Safety Promotion*, 27(4), 472–481.

Shekargoftar, A., Taghaddos, H., Azodi, A., Nekouvaght Tak, A., and Ghorab, K. (2022). "An integrated framework for operation and maintenance of gas utility pipeline using BIM, GIS, and AR." *Journal of Performance of Constructed Facilities*, 36(3), 04022023.

Shrestha, P. P., and Pradhananga, N. (2009). "GIS-based road maintenance management." *Computing in Civil Engineering*, 472–484.

Triana, E., and Labadie, J. W. (2012). "GIS-based decision support system for improved operations and efficiency conservation in large-scale irrigation systems." *Journal of Irrigation and Drainage Engineering*, 138(10), 857–867.

Tsilimantou, E., Delegou, E. T., Nikitakos, I. A., Ioannidis, C., and Moropoulou, A. (2020). "GIS and BIM as integrated digital environments for modeling and monitoring of historic buildings." *Applied Sciences*, 10(3), 1078.

Vacca, G., Quaquero, E., Pili, D., and Brandolini, M. (2018). "Integrating BIM and GIS data to support the management of large building stocks." *The International Archives of the Photogrammetry, Remote Sensing and Spatial Information Sciences*, 42, 647–653.

Valinejadshoubi, M., Moselhi, O., Bagchi, A., and Salem, A. (2021). "Development of an IoT and BIM-based automated alert system for thermal comfort monitoring in buildings." *Sustainable Cities and Society*, 66, 102602.

Wang, M., Deng, Y., Won, J., and Cheng, J. C. (2019). "An integrated underground utility management and decision support based on BIM and GIS." *Automation in Construction*, 107, 102931.

Welch, D. P., Sullivan, T. J., and Filiatrault, A. (2014). "Potential of building information modelling for seismic risk mitigation in buildings." *Bulletin of the New Zealand Society for Earthquake Engineering*, 47(4), 253–263.

Wong, M. S., Nichol, J., and Ng, E. (2011). "A study of the wall effect caused by proliferation of high-rise buildings using GIS techniques." *Landscape and urban planning*, 102(4), 245–253.

Yang, A., Han, M., Zeng, Q., and Sun, Y. (2021). "Adopting building information modeling (BIM) for the development of smart buildings: A review of enabling applications and challenges." *Advances in Civil Engineering*, 2021(1), 8811476.

Yang, S. W., Lee, Y., and Kim, S. A. (2025). "Design and validation of a real-time maintenance monitoring system using BIM and digital twin Integration." *Buildings*, 15(8), 1312.

Zerger, A., and Smith, D. I. (2003). "Impediments to using GIS for real-time disaster decision support." *Computers, Environment and Urban Systems*, 27(2), 123–141.

Zhang, J., Seet, B. C., and Tjing Lie, T. (2015). "Building information modelling for smart built environments." *Buildings*, 5(1), 100–115.

Zhu, J., Tan, Y., Wang, X., and Wu, P. (2021). "BIM/GIS integration for web GIS-based bridge management." *Annals of GIS*, 27(1), 99–109.

Chapter 8

The Demolition Stage

8.1 Learning Objectives

After completion of this chapter, readers will be able to:

- understand the use of GIS in the demolition stage,
- understand the use of GIS in demolition waste management, emergency management, and environmental impact assessment,
- understand the use of integration of BIM and GIS in the demolition stage, and
- understand the application areas of BIM and the integration of BIM and GIS in the demolition stage.

8.2 Introduction

Geographic information system (GIS) is used to develop the inventory of each existing infrastructure. For example, the inventory of where components or fixtures came from, the possibility of returning them to the manufacturers, and warranties for reuse. When an infrastructure reaches the end of its lifecycle or is near to its demolition stage, GIS helps in reducing demolition waste, resale, or planning for returning the usable components or parts. GIS is used to find the locations for dumping the demolition waste and the associated transportation costs. In the hilly regions, GIS is used to find the routes to transport demolition waste, considering the road gradient in case heavy vehicles are used. GIS is also used in identifying suitable regions for processing construction and demolition wastes. This chapter, in the following sections, covers the application areas of GIS, building information modeling (BIM), and the integration of BIM and GIS in the demolition stage of infrastructure projects.

8.3 GIS in the Demolition Stage

Demolition wastes threaten the environment due to poor disposal, limited landfill sites available, and continuously rising urbanization. GIS mainly supports waste quantification, recycling waste, landfill site selection, and monitoring of illegal

DOI: 10.1201/9781003759348-8

waste dumping. It aids in carbon emissions assessment, environmental impact assessments, and emergency management. GIS enhances damage assessment in case of floods, earthquakes, and dam failures, improving preparedness, mitigation, and resource allocations as discussed below:

8.3.1 Demolition Waste Management

A large amount of old construction and demolition waste resulting from both emergency and non-emergency events is generated and poses a great threat to the environment. Hence, understanding the total amount, composition, and rate at which waste is generated is important. GIS is used to quantify, formulate strategies to manage, and finally find locations for disposal of the old construction and demolition wastes (Wu et al. 2016). GIS is innovative in terms of systemization, visual representation, and quantifying old construction and demolition waste flows. It is also used to manage construction and demolition wastes in emergency situations, specifically in war-stricken environments (Madi and Srour 2019).

GIS is used to find suitable locations for construction and demolition waste dumping yards. The available landfill spaces are progressively declining, while the demand for quarry products is continuously increasing. The proper management and recycling of the waste help alleviate these problems. GIS is used to estimate quantities of construction and demolition wastes, decide suitable land for the construction of recycling facilities, and carry out an economic assessment of waste recycling.

The scarcity of raw materials and limited availability of landfill spaces due to the increasing rate of urbanization throughout the world have encouraged efforts to divert construction and demolition wastes away from landfills toward sorting, recycling, and reusing. GIS is used for developing an optimization-based decision-support system to select a cost and environmentally effective set of abandoned quarries to find locations for demolition waste landfills and processing facilities. GIS and integer linear programming are used for the development of waste management chains (Alzaghrini et al. 2019).

Paz et al. (2018) managed of construction and demolition waste in three stages: first is, to map construction and demolition wastes disposal points and classify the wastes according to their recyclability; second is, the development of a method for indicating suitable areas for installation of voluntary delivery locations for small waste generators; finally, the development of a method for indicating suitable locations for the installation of trans-shipment and waste sorting areas for large generators.

A huge amount of construction materials are required in urban areas for developing and maintaining buildings and other infrastructure. Hence, a huge amount of construction and demolition wastes are generated. Further, ageing building stocks also cause additional waste inflow. In order to assess urban metabolism with regard to buildings and other infrastructure, it is necessary to understand changes in its material accumulation both spatially and temporally. The material accumulation over time is elucidated using 4D GIS at an urban scale. The materials stock of buildings, roadways, and railways is estimated locally over time using a 4D GIS. The GIS-based database is used to find the spatial distribution of construction materials over time, estimate demolition waste based on characteristics of an area, and determine materials accumulation vertically, such as above and below the ground for recycling (Tanikawa and Hashimoto 2009).

8.3.2 Estimation of Carbon Emissions

Attention is needed to reduce carbon emissions from construction and demolition activities to meet the decarbonization target of the AEC industry. The carbon emissions accounting from the construction to the demolition stage is hardly done. GIS is used to develop building stock models, maintain statistical data, and maintain survey information used to create the spatiotemporal database for estimating the carbon emissions from the construction to the demolition stage (Bai et al. 2024). The logarithmic mean Divisia index approach is used to decompose spatiotemporal variations and identify the contributing factors for carbon emissions, focusing on the municipal level.

8.3.3 GIS in Environmental Impacts Assessments

Evaluating the potential health risks associated with air pollution near the demolition waste dumping yards is important. This makes it necessary to assess the exposure levels of nearby residents, workers, and sensitive populations such as children, aged groups, and individuals with respiratory sensitivity to pollutants. Another important parameter to assess the potential risks to human health is from contaminated water sources near the demolition waste dumping yards. GIS is used to analyze the spatial variations in air and water quality parameters around the construction and demolition recycling units (Chandrasekaran et al. 2023). The utilization of GIS for comprehensive analysis of waste management systems and in the identification of potential areas for their improvement provides valuable insights into the spatial distribution of air and water quality parameters around recycling units.

Construction and demolition waste, dumped illegally in ravines and open areas, contaminates soil and water and causes underground water pollution, soil contamination, and forest fires. The actual monitoring of illegal construction and demolition waste dumping and enforcing legislation against offenders is often difficult due to the large size of geographic areas. This needs to be monitored continuously; however, limited human and financial resources are available to environmental law enforcement agencies. GIS is used to identify areas under potentially elevated risks of illegal construction and demolition waste dumping. The locational factors that are significantly associated with the accumulated amount of waste on construction and demolition sites include distance to the nearest main road, depth of the ravine present at a site, and forest proximity (Seror and Portnov 2018). GIS is used to consider the locational factors and map the areas at high risk of illegal construction and demolition waste dumping for future monitoring. GIS facilitates environmental law enforcement authorities by helping them to focus on specific sites for inspection, save resources, and act against offenders more efficiently.

8.3.4 *Emergency Management*

Emergency management is comprehensive and complex that involving not just crisis-reactive responses to emergencies but also finding ways to avoid problems in the first place and preparing for those that undoubtedly will occur in the coming times. GIS is used in emergency management efforts for disaster preparedness, mitigation, response, and to summarize an ongoing effort to build a strong decision-support system. The database in GIS helps in emergency decision-making, preparedness, mitigation, and response efforts for the most common disasters like flooding, landslides, earthquakes, etc. GIS is used for better flood management by automating the task of determining the probable areas where floods may affect and integrating the results with other spatially distributed information systems for other solutions. GIS enables making more informed decisions before, during, and after a flood situation (Gunes and Kovel 2000). The database in GIS helps emergency management officers in decision-making, focusing on preparedness, mitigation, and response efforts. It enables emergency management officers to make better decisions. A well-organized response operation is required to mitigate chaotic situations during disasters. The efficient allocation of resources such as manpower, machinery, and suitable equipment is critical to perform disaster response operations. GIS is used to facilitate suitable equipment allocation under specific work conditions on affected sites in response to disasters (Chen et al. 2011).

8.3.5 *Damage Assessment*

Earthquakes have greater effects on society than most people think. These effects include loss of people, structural damage to infrastructure, economic impacts, and damage to fatalities/utilities. An earthquake lasts for a few seconds, and aftershocks continue for days, but the effect of damage does continue for years. The residential site safety and earthquake damage assessment studies play a crucial role in developing reliable rehabilitation and development programs, improving preparedness, and mitigating losses in urbanized areas. GIS is used for earthquake damage assessment and settlement planning (Hashemi and Alesheikh 2011). The estimation of seismic damage to buildings is complicated due to many sources of uncertainty. An artificial intelligence system and GIS are used to predict seismic damage to buildings at an urban scale, considering input uncertainties (Hansapinyo et al. 2020). It is possible to generate damage scenarios for earthquake-prone areas where only limited seismic data are available (Shin et al. 2020).

Mapping of areas that may be flooded due to the dam infrastructure failure is crucial. The dam-break hazard studies are lengthy, require large volumes of field data, and technical skills. GIS and hydrodynamic simulation are used to estimate the areas that may be flooded because of an instantaneous dam break due to its structural failure. GIS uses dam height, crest length, reservoir capacity, and a digital elevation model in the analysis for determining areas that may be flooded. It allows rapid dam-break hazard mapping by evaluating the dam-failure discharge hydrograph, propagation of flood in an area, and delineating flood-prone areas (Albano et al. 2019).

GIS is also used for estimating the potential loss of life from natural and dam failure floods. It is used to provide life-loss estimates for dam safety risk assessments. GIS is used to explore the options for reducing life-loss by minimizing the probability of dam failure or by improving the effectiveness of emergency planning and responses (Aboelata et al. 2003).

8.4 BIM in the Demolition Stage

BIM offers transformative capabilities for managing complexity, safety, and environmental impact during the demolition stage. It enables precise modeling of existing structures, helping engineers and demolition teams visualize hazardous zones, determine structural vulnerabilities, and develop safe deconstruction plans. In the demolition stage, when the infrastructure lifecycle is over, BIM contributes to reducing on-site risks, improving workers' safety, and minimizing potential secondary damage. BIM supports sustainable demolition practices by enabling detailed analysis of building materials and components, which is critical for effective waste management and resource recovery. By identifying materials suitable for reuse, recycling, or safe disposal, BIM significantly contributes to the reduction of demolition waste, supports economic practices, and lowers the carbon footprint associated with demolition activities. This aligns with broader sustainability goals and environmental regulations increasingly enforced in urban redevelopment projects. BIM integration with GIS, sensor networks, and real-time data feeds further enhances situational awareness and operational efficiency during emergency demolition or reconstruction efforts. This integrated approach enables quicker mobilization of resources and more resilient infrastructure planning in disaster-prone regions (Cheng et al. 2015).

Natural disasters, including landslides, wildfires, hurricanes, earthquakes, and floods, pose serious threats to both built infrastructure and the well-being of affected communities. In case of such challenges, the concept of disaster resilience has gained increasing importance. It focuses on enhancing the capacity of buildings to absorb shock, adapt to changing conditions, and recover swiftly in the aftermath of catastrophic events. BIM plays a pivotal role in supporting disaster resilience initiatives by facilitating informed decision-making throughout the lifecycle of the built environment (OpenAI 2025). Specifically, BIM aids in the development of resilient infrastructure by improving coordination and planning across both pre-disaster and post-disaster phases. Its applications span a wide array of critical areas, including supply chain optimization, accelerated construction and reconstruction processes, emergency response planning, and post-disaster rescue operations (Sertyesilisik 2017).

BIM is a critical enabler in the demolition and disaster recovery phases of infrastructure management. Its ability to provide data-rich visualizations and actionable insights allows stakeholders to optimize resource use, enhance safety, reduce environmental impact, and accelerate recovery. The following sections will delve deeper into these multi-dimensional applications, showcasing how BIM continues in the demolition stage:

8.4.1 BIM in Demolition Waste Management

Effective demolition waste management at the end is a critical component of sustainable lifecycle and urban development. A foundational aspect in this is quantitative waste prediction, which enables stakeholders to identify major sources of waste generation and formulate targeted waste control strategies. Accurate estimation of demolition waste volumes is essential not only for logistical planning and cost control but also for developing long-term waste management policies aligned with environmental sustainability goals. BIM has emerged as a powerful tool for supporting demolition waste estimation and management by integrating multi-disciplinary data into a centralized digital environment. BIM-based systems allow for the extraction of detailed material information, including types, quantities, and spatial distribution, directly from a 3D model of a building. This granular data enable the precise calculation of material volumes that will be generated as waste during demolition activities (Cheng and Ma 2013).

By identifying and categorizing materials before demolition, such as concrete, steel, wood, glass, etc., BIM facilitates more efficient coordination with waste recyclers and recovery firms. Recyclers are informed in advance about the types and volumes of reusable materials, enabling them to prepare accordingly and improve the effectiveness of recycling operations. BIM-based planning predicts waste collection logistics, including pickup requirements, sorting needs, and waste disposal fees based on material types, which enhances the economic feasibility and operational planning of demolition wastes. BIM also supports early-stage waste management planning. By integrating demolition waste estimation tools within the design phase of a building project, project teams can proactively evaluate end-of-life waste implications. Using BIM libraries and parametric modeling features, designers model material usage and predict demolition waste outcomes, ensuring that waste minimization strategies are embedded from the outset (Kim et al. 2017). This early intervention approach allows for more streamlined processing and sustainable outcomes throughout the building's lifecycle.

BIM contributes to overall process improvement in demolition waste management. Through digital coordination, real-time updates, and visualization capabilities, BIM enhances project transparency, reduces manual errors, and promotes the efficient segregation, handling, and minimization of demolition waste (Won and Cheng 2017). In typical demolition practices, waste is managed through three primary methods: landfilling, reuse, and recycling. Traditional documentation methods used to assess the environmental impact of these approaches tend to produce extensive paper-based records, requiring significant time and resources. BIM offers a more efficient alternative by enabling a digital assessment and comparison of the environmental consequences of each disposal method. BIM models are linked with lifecycle assessment tools to evaluate impacts such as resource depletion, biodiversity loss, and human health risks, thereby guiding stakeholders toward more environmentally responsible disposal methods (Wang et al. 2022).

BIM also plays a strategic role in addressing the challenges posed by urban road widening projects, where large-scale building demolitions are often required. Such projects lead to significant engineering, economic, and environmental disruptions. BIM, in conjunction with online mapping data and statistical analysis, is employed to estimate building demolitions without the need for physical site surveys (Jiang et al. 2022). By creating BIM models of existing roads and surrounding infrastructure, and comparing them with proposed road expansion designs, clash detection techniques are used to identify which buildings will be affected. This approach provides a cost-effective and practical solution for early-stage demolition estimation and urban planning, supporting data-driven decision-making and minimizing unnecessary site work.

8.4.2 *Waste Recycling and Reuse Planning*

Amid the growing global emphasis on environmental sustainability and green construction, the need for efficient construction and demolition waste management has become more pressing than ever. One of the persistent challenges in the AEC industry is the low rate of waste reuse and recycling, which contributes significantly to environmental degradation and inefficient resource utilization. Addressing this issue requires the adoption of advanced technologies and intelligent planning frameworks that can support more sustainable construction and demolition practices. BIM has emerged as a powerful enabler for improving construction and demolition waste management. As a data-rich digital platform, BIM facilitates the identification of reusable and recyclable building components by offering detailed insights into material specifications and quantities. The visualization and simulation capabilities of BIM allow stakeholders to evaluate the reuse potential of individual building elements even before demolition or renovation. This proactive approach improves material recovery and contributes to circular economy objectives.

A significant advancement in this domain involves the integration of temporal algorithms with 4D-BIM, which links 3D building models with construction schedules. 4D-BIM enables the simulation of material flows over time, allowing planners to optimize on-site waste segregation, plan reuse strategies, and reduce landfill dependency. This dynamic planning capability ensures that sequencing and material recovery processes are aligned, thereby improving overall resource efficiency and supporting sustainability goals (Guerra et al. 2020).

The combination of BIM with web-based technologies and cloud-based databases offers a scalable and automated solution for comprehensive construction and demolition waste management. These integrated systems facilitate real-time data collection, integration, mapping, analysis, and visualization, eliminating the need for manual tracking and improving operational accuracy. The adoption of such digital platforms enhances data interoperability across stakeholders and supports more transparent data-driven decision-making. In terms of logistics and traceability, the development of digital tracking systems provides a structured platform for managing key information, such as material inventories, reuse plans, and transportation logistics. These systems store and manage detailed datasets related to construction activities, material flow, and reuse strategies, offering reliable data support for project teams and regulatory authorities. Enhanced traceability also supports performance monitoring, compliance with environmental standards, and long-term reporting on sustainability metrics (Zheng et al. 2025). The integration of BIM with digital technologies not only increases the rate of material reuse and recycling but also contributes to more sustainable, efficient, and environmentally responsible construction practices.

8.4.3 Minimizing Carbon Emissions

The rapid pace of urban development and infrastructure renewal has led to a significant and continual rise in old building demolition waste, presenting a major environmental and logistical challenge for modern cities. The waste is not only generated in large volumes but is also dispersed across multiple sites, complicating its collection, transportation, and processing. A critical yet often underemphasized issue in this process is the high carbon footprint associated with demolition waste management. The heavy reliance on fuel-powered machinery for demolition and the extensive use of transportation vehicles contribute significantly to greenhouse gas emissions, particularly carbon dioxide. Identifying waste treatment methods that minimize these emissions remains a complex and site-specific challenge, especially considering the variations in waste composition, local recycling infrastructure, transportation distances, and energy sources used.

Traditionally, carbon emissions associated with demolition waste treatment have been inadequately addressed or entirely overlooked during the planning of the infrastructure lifecycle. BIM is now being recognized as a valuable tool to bridge this gap by enabling detailed and data-driven assessments of carbon emissions across the lifecycle of demolition waste. BIM models integrate geometric, material, and scheduling data, making it possible to simulate and analyze various demolition scenarios and their associated environmental impacts in a virtual environment. By incorporating lifecycle assessment tools within the BIM, project teams quantify and compare carbon emissions generated at different stages of waste management, ranging from on-site handling to off-site processing and final disposal. This capability supports the selection of low-emission strategies, such as prioritizing local recycling, reducing transport distances, or optimizing demolition sequences to minimize machine use.

Notably, on-site waste collection and sorting activities have been identified as the largest contributors to overall carbon emissions in the demolition waste stream, primarily due to the intensive use of diesel-powered equipment and inefficient material handling practices (Wang et al. 2018). By simulating these operations in a BIM-enabled framework, planners optimize equipment usage, reduce idling times, and identify opportunities for material reuse on-site, thereby lowering emissions and improving sustainability outcomes. Moreover, the environmental impact of recycling varies widely depending on the type of material, the method of processing, and the location of recycling facilities. BIM's ability to associate environmental data with specific building components allows for a granular analysis of these factors, enabling more accurate forecasting and smarter decision-making regarding material recovery and disposal. Leveraging BIM for the carbon assessment of demolition waste treatment marks a significant step toward environmentally responsible construction practices, decisions that align with broader climate goals, and sustainability mandates.

8.5 BIM and GIS Integration in the Demolition Stage

Rapid urbanization, population growth, and increasing demand for infrastructure put immense pressure on the use of natural resources. The AEC industry plays a vital role in shaping urban environments, which makes it one of the largest consumers of natural resources for raw materials. The industry faces critical challenges: constant demand for resources, excessive raw material consumption, and excessive construction and demolition waste generation. Current statistics suggest that the construction industry is responsible for nearly 50% of global solid waste, posing severe threats to ecosystems, public health, and climate goals. Although construction methods have matured over the years, sustainability practices, particularly in terms of waste reduction and energy conservation, are still evolving. Traditional construction practices in the AEC industry largely follow a linear "take–make–dispose" model, where raw materials are extracted, processed into products, used for a limited period, and then discarded as waste. This model not only depletes finite available resources but also results in illegal landfilling, environmental pollution, and high carbon emissions (McNeil-Ayuk and Jrade 2023).

To counter these drawbacks, the concept of circular economy has emerged as a transformative framework that redefines how materials are used in the entire lifecycle. Unlike the linear model, the circular economy emphasizes keeping resources in circulation for as long as possible through principles such as reducing, reusing, repairing, refurbishing, and recycling. The circular economy minimizes resource depletion and waste generation by keeping materials in circulation for as long as possible through reuse and recycling. In the AEC industry, this means designing infrastructure that can be easily disassembled, using materials with high recyclability, and incorporating secondary raw materials such as recycled concrete or reclaimed timber into new infrastructure projects. By doing so, the concept of circular economy reduces dependency on virgin materials, lowers energy consumption, and minimizes environmental impacts. The application of the circular economy in the AEC industry also provides significant economic opportunities. Recycled materials replace costly virgin raw materials, while innovative business models such as leasing or sharing construction equipment and materials promote efficiency and reduce overall costs. Moreover, the circular economy encourages the development of new industries and jobs in areas such as material recovery, recycling plants, green product innovation, etc. (Kovacic et al. 2020).

A crucial enabler of the circular economy in the AEC industry is the use of advanced digital technologies. In particular, the combination of BIM and GIS provides a powerful decision-support platform. BIM allows the development of highly detailed digital models that contain information about building materials, components, and systems across all lifecycle stages of an infrastructure. This enables precise tracking of material quantities, qualities, and potential for reuse or recycling. GIS, on the other hand, adds the spatial and geographical dimensions, contextualizing building information within broader urban environments. When combined, BIM and GIS provide a comprehensive digital representation of infrastructure stock, offering insights not only into what materials exist within structures but also where they are located and how they can be recovered efficiently (Ruiz et al. 2020). BIM and GIS integration enables tracking of material inventories over time, allowing stakeholders to predict when certain infrastructure reaches the end of its life cycle and plan for its reuse or recycling. BIM and GIS, together with the circular economy, may help in:

- Quantification of construction and demolition waste, where designers can estimate the amount of waste generated during construction and after demolition stages,
- Planning construction and demolition waste handling strategies where appropriate equipment, transportation routes, and storage facilities can be determined,
- Quantification of the economics of waste handling, recycling, and disposal of residual, and
- Design infrastructure with future material recovery, reuse, and recycling in mind, allowing for sustainable end-of-life cycle management.

The integration of BIM and GIS with the circular economy principles creates a foundation for sustainable lifecycle management of infrastructure. It enables stakeholders to design buildings with disassembly and reuse in mind, manage resources more effectively, and reduce waste generation during both construction and deconstruction stages. Furthermore, it supports policymakers and urban planners by offering accurate data for large-scale strategies on recycling facilities, material recovery, and sustainable urban development.

Lingering old or vacant facilities deteriorate the condition of an urban environment. The integration of BIM and GIS is used to match existing old or vacant facilities to the requirements of potential customers or owners to promote sustainable redevelopment and reuse (Zhang et al. 2009). BIM and GIS are used to automatically extract attributes from the virtual models and their surroundings for reusing old or vacant facilities. A profile is created for each existing vacant facility by combining BIM and GIS attributes that are matched to the desired BIM model that the aspiring users have provided. It promotes facility reuse and presents the most suitable vacant facilities to users (Wang et al. 2016). BIM and GIS are also used to identify suitable locations for redevelopment in an area. BIM and GIS together are used for the development of detailed 3D building models along with their surroundings which are used for assessment and visualization of flood damage to buildings (Amirebrahimi et al. 2016).

8.6 Conclusions

A large amount of demolition waste is generated due to rapid urbanization. It is imperative to understand the amount, composition, and flow of the generated waste. GIS is used to quantify the demolition waste from generation to final disposal location and formulate the strategies for its management. The disposal of demolition waste results in many environmental-related issues; however, there is a serious lack of its environmental impact assessment, in which GIS has innovative solutions. The integration of GIS and Internet of Things technology has revolutionized the waste management system by providing real-time data and insights to decision-makers in formulating policies.

BIM plays a critical role in the demolition stage by accurately modeling existing infrastructure, supporting hazard identification, improving deconstruction planning, reducing risks, and protecting workers. Through detailed materials data, BIM enhances demolition waste prediction, recycling, and reuse, promoting circular economy practices and sustainable outcomes. BIM integrated with GIS and digital platforms optimizes waste logistics, resource recovery, and carbon emission reduction. In disaster resilience, BIM supports risk assessment, emergency planning, and reconstruction by enabling real-time coordination and data-driven decisions, ensuring safer, greener, and more resilient infrastructure development. BIM-GIS integration supports sustainable urban redevelopment by matching vacant facilities with user needs through attribute extraction and profiling. Integration aids facility reuse, redevelopment sites, and visualization of existing building conditions. Detailed 3D models of an infrastructure along with their surroundings in GIS, enable flood damage assessment and enhance decision-making for adaptive reuse and urban resilience.

8.7 Exercises

Question 8.1: How does the inventory of components and fixtures of existing infrastructure in GIS help during the demolition stage?

Question 8.2: In what ways does GIS contribute to reducing demolition waste during the end-of-life stage of infrastructure projects?

Question 8.3: How is GIS used in finding suitable locations for the dumping of demolition waste and estimating its transportation costs?

Question 8.4: How does GIS assist in emergencies, particularly in war-stricken environments, for demolition waste management?

Question 8.5: What role does GIS play in estimating demolition waste quantities and selecting land for establishing recycling facilities?

Question 8.6: How does GIS help in estimating and analyzing carbon emissions during the demolition stage of infrastructure projects?

Question 8.7: In what ways does GIS support environmental impact assessments related to air and water pollution near demolition waste recycling sites?

Question 8.8: How does GIS facilitate in monitoring of illegal demolition waste dumping in ravines, forests, and open areas?

Question 8.9: What is the role of GIS in emergency management and disaster response planning?

Question 8.10: How does GIS contribute to earthquake damage assessment and re-settlement planning?

Question 8.11: In dam-break hazard studies, how does GIS support rapid hazard mapping and flood-prone area delineation?

Question 8.12: In what ways does BIM contribute to waste recycling, reuse planning, and minimizing carbon emissions during the demolition stage?

References

Aboelata, M., Bowles, D. S., and McClelland, D. M. (2003). "GIS model for estimating dam failure life loss." *In: Proceedings of the 10th engineering foundation conference on Risk-Based Decision-making, Santa Barbara*, 126–145.

Albano, R., Mancusi, L., Adamowski, J., Cantisani, A., and Sole, A. (2019). "A GIS tool for mapping dam-break flood hazards in Italy." *ISPRS International Journal of Geo-Information*, 8(6), 250.

Alzaghrini, N., Srour, F. J., and Srour, I. (2019). "Using GIS and optimization to manage construction and demolition waste: The case of abandoned quarries in Lebanon." *Waste Management*, 95, 139–149.

Amirebrahimi, S., Rajabifard, A., Mendis, P., and Ngo, T. (2016). "A framework for a micro-scale flood damage assessment and visualization for a building using BIM–GIS integration." *International Journal of Digital Earth*, 9(4), 363–386.

Bai, Y., Yamashita, N., Guo, J., Nagata, S., Shirakawa, H., and Tanikawa, H. (2024). "Dynamic analysis of carbon emissions from construction and demolition activities in Japan: Revealed by high-resolution 4D-GIS modelling." *Journal of Environmental Management*, 360, 121099.

Chandrasekaran, H., Subramani, S. E., Partheeban, P., and Sridhar, M. (2023). "IoT-and GIS-based environmental impact assessment of construction and demolition waste dump yards." *Sustainability*, 15(17), 13013.

Chen, A. Y., Mora, F. P., and Ouyang, Y. (2011). "A collaborative GIS framework to support equipment distribution for civil engineering disaster response operations." *Automation in Construction*, 20(5), 637–648.

Cheng, H. M., Yang, W. B., and Yen, Y. N. (2015). "BIM is applied in historical building documentation and refurbishment." *The International Archives of the Photogrammetry, Remote Sensing and Spatial Information Sciences*, 40, 85–90.

Cheng, J. C., and Ma, L. Y. (2013). "A BIM-based system for demolition and renovation waste estimation and planning." *Waste Management*, 33(6), 1539–1551.

Guerra, B. C., Leite, F., and Faust, K. M. (2020). "4D-BIM to enhance construction waste reuse and recycle planning: Case studies on concrete and drywall waste streams." *Waste Management*, 116, 79–90.

Gunes, A. E., and Kovel, J. P. (2000). "Using GIS in emergency management operations." *Journal of Urban Planning and Development*, 126(3), 136–149.

Hansapinyo, C., Latcharote, P., and Limkatanyu, S. (2020). "Seismic building damage prediction from GIS-based building data using artificial intelligence system." *Frontiers in Built Environment*, 6, 576919.

Hashemi, M., and Alesheikh, A. A. (2011). "A GIS-based earthquake damage assessment and settlement methodology." *Soil Dynamics and Earthquake Engineering*, 31(11), 1607–1617.

Jiang, F., Ma, L., Broyd, T., Chen, K., Luo, H., and Du, M. (2022). "Building demolition estimation in urban road widening projects using as-is BIM models." *Automation in Construction*, 144, 104601.

Kim, Y. C., Hong, W. H., Park, J. W., and Cha, G. W. (2017). "An estimation framework for building information modeling (BIM)-based demolition waste by type." *Waste Management and Research*, 35(12), 1285–1295.

Kovacic, I., Honic, M., and Sreckovic, M. (2020). "Digital platform for circular economy in AEC industry." *Engineering Project Organization Journal*, 9(1), 16.

Madi, N., and Srour, I. (2019). "Managing emergency construction and demolition waste in Syria using GIS." *Resources, Conservation and Recycling*, 141, 163–175.

McNeil-Ayuk, N., and Jrade, A. (2023). "Integrating building information modeling (BIM), GIS, and circular economy (CE) for the construction and deconstruction waste based on construction methods at the conceptual design stage of buildings." In *Canadian Society of Civil Engineering Annual Conference*, Cham: Springer Nature Switzerland, 343–356.

OpenAI (2025). ChatGPT (Version 5.1) [Large language model]. https://chat.openai.com/

Paz, D. H. F. D., Lafayette, K. P. V., and Sobral, M. D. C. (2018). "GIS-based planning system for managing the flow of construction and demolition waste in Brazil." *Waste Management and Research*, 36(6), 541–549.

Ruiz, L. A. L., Ramón, X. R., and Domingo, S. G. (2020). "The circular economy in the construction and demolition waste sector–A review and an integrative model approach." *Journal of Cleaner Production*, 248, 119238.

Seror, N., and Portnov, B. A. (2018). "Identifying areas under potential risk of illegal construction and demolition waste dumping using GIS tools." *Waste Management*, 75, 22–29.

Sertyesilisik, B. (2017). "Building information modeling as tool for enhancing disaster resilience of the construction industry." *Transactions of the VSB - Technical University of Ostrava*, XII(1), 9–18.

Shin, J., Choi, H. S., and You, Y. C. (2020). "BIM/GIS platform-based seismic damage assessment system for building structures." *Journal of the Korean Association for Spatial Structures*, 20(1), 18–22.

Tanikawa, H., and Hashimoto, S. (2009). "Urban stock over time: Spatial material stock analysis using 4D-GIS." *Building Research and Information*, 37(5-6), 483–502.

Wang, J., Wei, J., Liu, Z., Huang, C., and Du, X. (2022). "Life cycle assessment of building demolition waste based on building information modeling." *Resources, Conservation and Recycling*, 178, 106095.

Wang, J., Wu, H., Duan, H., Zillante, G., Zuo, J., and Yuan, H. (2018). "Combining life cycle assessment and building information modelling to account for carbon emission of building demolition waste: A case study." *Journal of Cleaner Production*, 172, 3154–3166.

Wang, T., Krijnen, T., and De Vries, B. (2016). "Combining GIS and BIM for facility reuse: A profiling approach." *Research in Urbanism Series*, 4, 185–204.

Won, J., and Cheng, J. C. (2017). "Identifying potential opportunities of building information modeling for construction and demolition waste management and minimization." *Automation in Construction*, 79, 3–18.

Wu, H., Wang, J., Duan, H., Ouyang, L., Huang, W., and Zuo, J. (2016). "An innovative approach to managing demolition waste via GIS (geographic information system): A case study in Shenzhen City, China." *Journal of Cleaner Production*, 112, 494–503.

Zhang, X., Arayici, Y., Wu, S., Abbott, C., and Aouad, G. (2009). "Integrating BIM and GIS for large-scale facilities asset management: A critical review." *The Twelfth International Conference on Civil, Structural and Environmental Engineering Computing*, Funchal, Madeira, Portugal.

Zheng, B., Yang, Y., Chi, H. L., Luk, C., and Chan, A. (2025). "Development of a BIM-based application for visualizing the reuse and recycling potential of modular building systems." *Journal of Building Engineering*, 105, 112500.

Chapter 9

Research Directions

9.1 Learning Objectives

After the completion of this chapter, readers will be able to:

- understand the future areas of research required to enhance the use of GIS in the different lifecycle stages of infrastructure projects,
- understand how improvements in technology and interoperability between BIM and GIS can significantly improve the application areas of GIS in the different lifecycle stages of infrastructure projects,
- understand the intervention of digital technologies in the AEC industry and the need for technological research priorities, and
- understand how the age, experience, and education level of professionals in the AEC industry are the factors that influence the use of GIS in the different lifecycle stages of infrastructure projects.

9.2 Introduction

The lifecycle stages of an infrastructure project use detailed project reports, architectural and structural drawings, blueprints, cost estimates, site layouts, execution schedule, specifications, operating procedures, etc., which are developed and maintained separately by using different software tools by professionals in the architecture, engineering, and construction (AEC) industry. The AEC industry professionals have to go through such information throughout the lifecycle stages but are difficult to bridge together. Owners, architects, engineers, contractors, operators, and users face challenges due to the variety of information formats used for such graphical and non-graphical information about an infrastructure. This also creates an interoperability problem. If both graphical and non-graphical information are developed, analyzed, and managed in a single environment and changes are made to such information at one place only, the information will be consistent with each other. In the AEC industry, building information modeling (BIM) provides capabilities to solve this problem, which involves the bridging, analysis, and use of large volumes of both graphical and non-graphical information. BIM

DOI: 10.1201/9781003759348-9

provides strong premises to overcome the fragmented nature of the AEC industry. BIM facilitates information exchange and reduces interoperability problems during the entire lifecycle stages. However, in BIM, the main focus is on indoor planning tasks such as 3D modeling, calculation of work quantities, space use, energy consumption, like heating, ventilation, air conditioning, etc. BIM is mainly targeting the building sector or vertical infrastructure only; future research is required to explore the use of information modeling for all types of infrastructure projects.

BIM lacks in geospatial analyses like site selection, route planning to a site, planning for delivery of construction materials and services to a site, supply chain management, construction site layout planning, site operation planning, emergency evacuation operations, etc. Therefore, geospatial analysis capabilities provided by geographic information system (GIS) have their own strengths as discussed in the previous chapters. Even after the invention of BIM, the AEC industry is looking for geospatial analysis capabilities available in GIS. The previous chapters cover the application areas of GIS in different lifecycle stages of infrastructure projects in which BIM is lacking. Different types of ground data required in the planning and management of the lifecycle of large infrastructure projects like residential colonies, industries, hydroelectric projects, etc., are already being represented in GIS.

BIM mainly focuses on indoor modeling and analysis, while GIS, on the other hand, facilitates the planning of outdoor tasks. BIM is used to model an infrastructure that does not exist. BIM is used for representing the maximum level of detail in terms of geometry and attributes of an infrastructure to be developed. On the other hand, GIS is used to represent surroundings that already exist. Hence, GIS data models represent surroundings in the most abstract way in terms of geometry; however, locations and external geometry of different features on the ground play a major role in the planning process. Hence, GIS has less focus on indoor drafting capabilities and more on surface modeling that facilitates spatial analysis related to relief, including slope, aspect, elevation, visibility, cut and fill volume calculation, surface area, 3D visualization of a surface, perspective views, etc., as covered in Chapter 2. Therefore, the use of BIM and GIS together facilitates understanding the effects of surroundings on each lifecycle stage of an infrastructure project and the impacts on surroundings because of new infrastructure additions. Therefore, to understand both, the new infrastructure to be developed and its surroundings, research on the integration of BIM and GIS is required.

The AEC industry professionals must be aware of the capabilities of GIS, its application areas in the lifecycle stages of infrastructure projects, and the future research areas required for extending its use in the industry (Bansal 2012). Chapter 2 makes the AEC professionals aware of GIS, Chapters 3 to 8 provide the application areas of GIS in the lifecycle stages of infrastructure projects, and the present chapter provides future research areas for extending the use of GIS in the AEC industry. There are certain functional overlaps between BIM and GIS in the area of spatial information analysis; on the other hand, certain key differences also exist (Bansal 2011).

In addition to BIM and GIS, the use of digital tools, like artificial intelligence, large language models, ChatGPT, etc., in the AEC industry has the potential to address persistent challenges such as skill shortages, inefficiencies in education, training, and professional practice. By embedding artificial intelligence mainly in learning and operational workflows, the industry can bridge the gap between theory and practice. It will also accelerate innovation and productivity in the industry. Artificial intelligence tools foster interdisciplinary learning environments where theoretical knowledge is reinforced through hands-on and practical experience. For instance, immersive simulations can train workers in safety protocols or construction techniques, and predictive analytics can anticipate resource demands, equipment needs, and workforce allocation (Jelodar 2025). These capabilities enhance workforce readiness and create smarter and more adaptive construction processes. For ensuring sustainability, the AEC industry must establish robust governance frameworks, ethical standards, and equitable access to artificial intelligence tools. Overcoming these barriers requires developing modular and scalable artificial intelligence solutions tailored to different contexts within the AEC industry (Akhimien et al. 2021). The following sections provide the areas for future research to extend the use of GIS and digital technologies in the different lifecycle stages of infrastructure projects:

9.3 Research Directions to Extend the Use of GIS

The research directions to extend the use of GIS in different lifecycle stages of infrastructure projects are provided in the following sections:

9.3.1 *In Feasibility Stage*

GIS has wide application areas like database management, modeling of surroundings, modeling of surface and sub-surface facilities/utilities, modeling of site conditions and constraints, modeling of spatial relationships, selection of sustainable site, planning site access, and extending services to a site in the feasibility stage. GIS provides a set of tools to locate an infrastructure, keeping the sustainability of a site in consideration. The virtual model of an infrastructure, along with its surroundings, helps in evaluating different alternatives and finding the most suitable site in the feasibility stage. However, a few areas listed in the following subsections still require further future considerations:

9.3.1.1 3D Modeling Capabilities

3D modeling capabilities developed in GIS are not available to the extent available in BIM software tools, which are widely used in the AEC industry. GIS software tools sometimes use a 2.5D approach to represent 3D objects on the surface of the earth. GIS software tools offer very limited 3D modeling capabilities for the modeling of components of an infrastructure. Hence, researchers and developers have the challenging role of maturing the 3D modeling capabilities in GIS software tools. The researchers and developers have to show the GIS users the possibilities and constraints of 3D GIS in order to obtain a serious breakthrough in 3D GIS. This will facilitate the simulation of the lifecycle stages of an infrastructure project along with its surroundings to make lifecycle planning and management more efficient and realistic.

9.3.1.2 3D Analysis Capabilities

GIS facilitates the calculation of 3D surface areas, projected areas, volume between two surfaces, visibility analysis, etc. Though many of the commercially available GIS software tools offer spatial analysis in two-dimensional (2D), however is very little in 3D. The main reason behind this is that database management system (DBMS) used in GIS software tools does not support 3D data types as these work with points, lines, and polygons features. Therefore, future research is required to develop functionality for 3D spatial analysis. It will enable 3D geospatial analysis in the lifecycle stages of infrastructure projects. The advancements in 3D analysis will facilitate the extraction of partial 3D models of an infrastructure that fulfill 3D spatial queries or constraints. There is a strong need for new user-friendly approaches to analyze complex 3D components along with their surroundings in GIS.

9.3.1.3 3D Topology

The topology in GIS enables the modeling of spatial relationships and performs analyses such as what infrastructure component is next to another, what component is within another, which components intersect each other, etc. Topology is a major advantage of using GIS over the existing BIM-based software tools in the AEC industry. Most of the GIS software tools construct 2D topological models and lack in 3D. The complexity of the topological relationships increases from 2D to 3D, thus, requiring new approaches for 3D topological modeling in GIS. There are a variety of topological data models in GIS that the AEC industry commonly needs. These models may contain one-dimensional topologies like networks (for example, roads), 2D topologies like areas (for example, working areas and layouts), or a combination of two (for example, a database containing an access road network, locations, and a construction site). However, research and development are still required in the 3D topology for better planning and management of the feasibility stage of an infrastructure (Borrmann and Rank 2009).

9.3.1.4 Interoperability Among Different Disciplines

The information sharing among different lifecycle stages of an infrastructure project needs to be seamless between planners, designers, operators, users, environmentalists, ecologists, etc. The use of GIS has not been explored for different types of infrastructure projects and completely for different lifecycle stages of an infrastructure project. The information gathered or generated in one stage of an infrastructure project or by one team does not transfer seamlessly to other stages or to other teams due to the lack of interoperability. This requires different information models and modeling capabilities in GIS. To meet such requirements more detailed research is required in assessing the suitability of particular data models (vectors, raster, or some other) in GIS. Object-oriented modeling in GIS could be an alternative to represent components of an infrastructure and modeling spatial relationships among them for supporting different lifecycle stages of infrastructure projects.

9.3.2 In Design Stage

GIS has application areas like public participation in the planning and design of infrastructure projects, construction quantity takeoffs, cost estimation, and sunlight visualization in the design stage. However, a few areas listed below still require future consideration:

9.3.2.1 Interoperability Between BIM and GIS

The Open Geospatial Consortium (OGC) with other standards organizations and industry associations develops open geospatial standards to support productivity across the supply chains of building design, physical infrastructure, capital projects, and facilities management industries. OGC is continuously practicing for implementing standards for publishing and streaming 3D geospatial assets related to urban planning and landscape visualization. Due to the lack of a common digital language of open software interface and information encoding standards, billions of dollars are spent on converting, recreating, and searching for infrastructure-related data. To improve communication among different lifecycle stages of an infrastructure project, the suitability of data exchange standards needs to be investigated up to the implementation level.

The mission of OGC is to deliver interface specifications for geographically referenced data. OGC specifications apply to generic spatial entities such as points, lines, and polygons and need to be contextualized before they are used for a particular application. OGC has developed several specifications to describe, encode, and transport information pertaining to spatial entities. Although, OGC is dealing with interoperability issues in the geographically referenced data but it is important for the AEC industry to understand its own interoperability problems. The AEC industry, through interactions with GIS professionals may become aware of how the GIS industry perceives interoperability problems that need to be addressed and resolved in the AEC industry also if the use of GIS is to be promoted in the design stage.

On the other hand, industry foundation classes (IFC) are a common building information model developed for the AEC industry by the International Alliance for Interoperability. IFC specifications are open and available. These are neutral and open specifications that are not controlled by a single vendor or group of vendors. OGC and IFC are living their separate lives, making their integration as one of the important research areas. The researchers and software developers may become aware of how the AEC industry perceives their data, difficulties, interoperability problems that need to be addressed, and resolved through interactions with professionals using BIM and GIS. It is also important for the AEC industry to begin to understand its own interoperability problems and look for solutions (Zhu et al. 2019).

The existing commercial and non-commercial solutions for interoperability problems between BIM and GIS mainly focus on developing data standards. The AEC industry and geospatial community are developing standards to enable information exchange between BIM and GIS. However, the interoperability problem may be beyond the data standards which require further investigation. The use of Open BIM software tools in the AEC industry is limited because open standards are targeting the building sector only, on the other hand, Open GIS software tools mainly rely on standards that need to be brought together using future research.

9.3.2.2 Construction Cost Estimates

GIS-based construction cost estimation works well for repetitive tasks or activities. The cost calculated is based on the data in resource tables which need to be updated on the basis of past practices. The accuracy of the cost estimation depends upon the correctness of the data in the tables and dimensions of components of an infrastructure in different data layers. However, the calculation of the construction cost of infrastructure like buildings still requires further research. Research is required for the calculation of location-based construction cost estimates and production rates.

9.3.3 In Planning Stage

The use of GIS is reported for the calculation of site-specific rates of different items of work, planning and scheduling of repetitive and non-repetitive infrastructure construction projects, location-based planning and scheduling, 4D modeling, resolving workspace conflicts, planning for safety during construction, and construction visualization in the planning stage. The virtual model of an infrastructure, along with its surroundings, helps in the planning and scheduling of infrastructure projects in a much better way. However, a few areas listed in the following subsections still require future investigations:

9.3.3.1 Construction Planning and Scheduling

The execution schedule of an infrastructure construction project cannot be finalized without considering available space, site topography, and site constraints. The schedule generated using commercially available planning and scheduling software tools does not consider the available space, site topography, and site constraints. The capability of a planner to mentally visualize the available site space, site topography, and site constraints to develop an execution schedule varies depending on the experience, knowledge, and individual perspective of a planner. Hence, effectiveness in planning varies from planner to planner. GIS helps in understanding the site conditions, surrounding topography, transportation routes, etc., to facilitate the generation of an executable schedule.

The construction schedule generated in the commercially available scheduling tools may be used in GIS but cannot be corrected or updated. The capabilities of the scheduling tool developed in GIS are not comparable to the functionality available in commercially available scheduling tools. The development of full scheduling functionality in GIS-based scheduling tools is required or an application-level integration of the commercially available scheduling tools with GIS is required. GIS-based scheduling tool uses activity-on-arrow representation in which activity splitting is not considered; therefore, this requires further investigations (Bansal and Pal 2008).

9.3.3.2 4D Modeling of Execution Sequence

The interpretation of the execution schedule without any link of its activities with the graphical aspects is difficult. This makes a schedule difficult to check for its completeness, congestion, and logical errors. 4D modeling of the execution sequence increases the comprehensibility of a project schedule and allows users to detect potential problems prior to the construction stage. However, the construction plan is affected by or has an impact on existing public facilities/utilities in its immediate surroundings, which are not considered in the simple 4D modeling software tools. GIS helps in demonstrating the available space, site topography, site constraints, immediate surroundings, etc. The schedule generated in commercially available software tools is also used in GIS to generate 4D models and literature also suggests the use of GIS-based scheduling tools (Bansal and Pal 2008). However, the full automation of the linking of 3D components with various activities in the schedule in the GIS environment is required to develop a 4D execution sequence in GIS.

9.3.3.3 Operational-Level Visualization

4D GIS depicts the evolution of an infrastructure virtually, not the interaction among the resources that build it. It is not possible to visualize the execution sequence at the operational level that develops an infrastructure in GIS-based 4D modeling. Operational-level visualization differs significantly in concept, content, and usage when compared to 4D modeling. The operational-level visualization is used to check the executability of the schedule. Executability is checked with respect to the surroundings of an infrastructure. The operational-level visualization provides a key to significant savings in project time duration, costs, and the reduction of several unwanted problems. To visualize construction at the operational level in GIS it is necessary to model the working of equipment and workers, handling of materials, construction and dismantling of TFs, and modeling of complex operations in addition to the 4D modeling of an infrastructure. Viewing interaction among different crews and/or equipment is part of operational-level visualization to meet the project schedule. There is a need for operational-level visualization in GIS to fully explore a construction plan for betterment. Working in a well-organized operational environment with a minimum amount of time spent on resolving workspace conflicts on a site is an important factor in meeting the project schedule. Hence, more research on operational-level visualization in GIS is required for better decision-making.

9.3.3.4 *Sustainable Construction Planning*

Due to the continuous increase in the population of the world, the demand for planning for the sustainable lifecycle of infrastructure projects is increasing continuously. The people involved in the planning of different lifecycle stages need software tools to evaluate the environmental impacts of infrastructure projects on urban microclimate and outdoor comfort. GIS is a tool for sustainable planning and makes it possible to have a better understanding of the surrounding environment. GIS also helps urban designers in simulating and evaluating the environmental impacts of their projects. Recognizing the importance of GIS, it is necessary to explore the possibility of checking each change in surroundings for its sustainability perspective. Ever-changing priorities of sustainable planning, development, and minimizing environmental impacts increase complexity, which requires further research.

9.3.4 *In Construction Stage*

The use of GIS is reported for screening of contractors, planning human workforce, materials trading, site layout planning, earthwork planning, site drainage planning, identification of locations for setting heavy equipment on a site, reducing construction wastage, and tracking site dynamics in the construction stage. The virtual model of an infrastructure along with its surroundings, helps in the construction stage. However, a few areas listed in the following sub-sections still require future consideration:

9.3.4.1 *Construction Site Planning*

Most of the GIS-based analyses are 2D, therefore, works done in the areas of developing site layouts, materials layouts, and predicting the space covered by equipment at the site are based on 2D geospatial analyses. Hence, good user-friendly 3D geospatial analysis capabilities need to be developed for various types of analyses required in the construction planning stage. Improvements are needed in GIS-based site layout planning to include permanent facilities in 3D, existing electric/water/sewer networks, and equipment positions which are generally not considered.

9.3.4.2 *Sustainable Construction*

Ever-changing priorities of sustainable development and minimizing environmental impacts due to construction increase complexity which requires further research. This process has not been explored much because of the large volume of information needed and the complicated interaction between many of the involved resources and surroundings. Planning and implementation of sustainable construction practices often require complex geospatial analysis. Although the construction of environmentally sustainable infrastructure might include the use of innovative construction materials, recycling, reuse, and pre-construction consultation with ecologists and environmentalists but understanding the interaction between a natural ecosystem and an infrastructure is also essential.

9.3.4.3 *Web/Server-Based GIS in Construction*

Fragmentation is an inherent characteristic of the AEC industry. It is a serious obstacle to communication and collaboration among people involved in the construction stages. Collaboration is essential to generate and timely sharing of updated information among people involved. Web/server-based GIS has made it much easier for the AEC industry to collect information in a central repository and share it over the Internet regardless of physical locations. An essential element of internet-based GIS for infrastructure projects is information management which provides the right information to the right person at the right time. The multiple persons working on infrastructure projects may be provided user-friendly server/web-based GIS platform for information sharing and analysis. It is easily possible to share the geospatial information of an infrastructure project on the web and use that where geospatial operations are carried out. For example, cloud computing relates a building to its geographical location, planners at different geographical locations work on the same project together in real time without consuming vast resources (Akıncı et al. 2010).

9.3.5 *In Operation and Maintenance Stage*

Generally, the operation and maintenance programs are designed considering an infrastructure in isolation. The surroundings of an infrastructure are considered manually by the professionals during the actual execution of programs. This results in different opinions about design and execution. This generates the need for a virtual model of an infrastructure and its immediate surroundings in GIS during the O&M stage. It is also used for site drainage planning in case of heavy rainfalls, cloud bursts, or flooding of an infrastructure and its surroundings.

There is a need to develop a central database to make reliable and accurate decisions regarding road maintenance and management considering traffic counts, pavement conditions, maintenance costs, and alternative roads available. The central database must be used to justify a budgetary request for the investment of public funds. Future research may focus on how to develop and provide a reliable database for O&M decision-making and other online services from the database.

GIS and artificial intelligence techniques together may be used to develop an intelligent O&M management system for optimizing maintenance programs. Artificial intelligence uses human-like reasoning concepts to optimize O&M management programs and automatically detects faults, components that need repairs, and evaluates cracks from images of buildings, roads, etc. It will be helpful in identifying high-risk road sections, defective components of infrastructure, and support effective communication for ensuring the safety of the infrastructure. Future research may explore how to use intelligent technology or artificial intelligence to efficiently and intelligently develop O&M management systems for large infrastructures like dams.

The health of an infrastructure is related to the safety and health of its users. The development of a suitable scientific evaluation system for infrastructure health monitoring may avoid the consequences of infrastructure failure. This requires future research investigations. The integration of BIM and GIS is necessary for integrating the internal environment of an infrastructure with the street or road networks for designing optimal routes in case of emergency evacuation during urban disasters. Therefore, the maximum number of multistory buildings in an urban area may be virtually modeled and linked to its surrounding networks for emergency planning.

9.3.6 In Demolition Stage

A large amount of old construction and demolition waste is generated and poses a great threat to the environment. The available landfill spaces for such waste are progressively declining, while the demand for spaces is continuously increasing. The capabilities of GIS may be enhanced for practitioners to estimate the quantities of waste and decide suitable sites for landfill and recycling facilities. GIS must be used to map the spatial distribution of construction waste generation over time and decide suitable sites for landfill and recycling facilities at the city or district level. GIS must be used to model the spatial variations in air and water quality parameters around every construction and demolition landfill site or recycling unit.

The waste dumped illegally in open areas contaminates soil and causes air and water pollution. The actual monitoring of illegal waste dumping and enforcing legislation against offenders over large geographic areas may be possible with the help of GIS. It will facilitate the environmental law enforcement authorities by helping them to focus on site-specific inspection, save resources, and act against offenders more efficiently. This requires future research for its actual implementation. There is a need to develop a central database in GIS that will help in emergency decision-making, preparedness, mitigation, and response efforts for the most common disasters like flooding, landslides, earthquakes, etc. GIS is used to facilitate equipment allocation under specific work conditions on sites in response to disasters.

BIM and GIS are used to automatically extract attributes from the virtual model of an infrastructure and its surroundings for its repair and maintenance. It promotes infrastructure reuse and presents the most suitable facilities to users. This has been demonstrated in many studies, however, to provide this technology to stakeholders, it requires future research and investigation.

9.4 Technological Intervention

Earlier, infrastructure development was manual, referred to as Construction 1.0, a period known as the craftsmanship era. This era preceded the mechanization period, during which manual labor, craftsmanship, and traditional methods were employed. The main materials used in construction during this era were stones, wood, bricks, lime mortar, etc. The tools used in this era were hammers, chisels, pulleys, etc. The construction knowledge was transferred through laborers, masons, and builders. This resulted in low productivity and high dependency on human skills.

Electricity came into existence, and with the Industrial Revolution, steam and later internal combustion engines were introduced. Mass production of new construction materials started. The revolution introduced new construction materials, like steel, cement, prefabricated members, etc., and machines, like cranes, concrete mixers, bulldozers, etc. This was the mechanization and industrialization era known as Construction 2.0. Standardization was also introduced in the industry through the development of codes/standards for engineering practices. For better accuracy and efficiency as compared to manual paper-based drawings, CAD was introduced for digital drafting. However, the AEC industry remained fragmented because of low data support and dependency on many manual working methods.

Information technology was introduced in the industry; it was the digital and BIM era known as Construction 3.0. Mainly, BIM, GIS, and project management software tools were introduced. BIM was used for the development of virtual 3D models with data-rich components for collaboration across different lifecycle stages. GIS was used to have a site-specific context of infrastructure. The problems faced in this era were partial digitalization, difficult BIM and GIS integration, limited interoperability across different disciplines, resistance to the adoption of new technologies, and data silos.

In the next stage, full-scale digital transformation of the AEC industry is being achieved through the adoption of BIM, GIS, artificial intelligence, Internet of Things, digital twins, robotics, drones, blockchain, smart sensors, augmented reality/virtual reality, 3D printing, etc. It is the digital transformation in construction, where advanced technologies are used in every stage of the project lifecycle. It is the smart and automated era known as Construction 4.0. It is the latest stage in the evolution of the construction industry, inspired by Industry 4.0. The main objectives to be achieved are high productivity, efficiency, low cost, minimum construction time, net-zero emissions, sustainability, and integration of different lifecycle stages. The main focus is on sustainability through the concept of circular economy practices. The digital and physical technologies are merged to make construction smarter, faster, safer, and more sustainable (OpenAI 2025). It provides smart and data-driven infrastructure, focusing on lifecycle integration where data flows seamlessly through all lifecycle stages. The following are the technologies used in Construction 4.0:

- **BIM:** is used for collaborative design, it helps in developing data-rich 3D models which are used for design, cost estimation, scheduling, sustainability analysis, clash detection, execution, lifecycle management, etc., as discussed in earlier chapters,
- **GIS:** supports infrastructure planning, site selection, and integration with BIM for providing site-specific context as discussed in earlier chapters,
- **Digital twin:** is a virtual, computer-generated replica of a physical object, process, or system that is updated with real-time data obtained from different sensors, used in continuous simulation, analysis, and monitoring. The digital twin models are used to understand, optimize, and predict the behavior of their real-world counterparts across their entire lifecycle, enabling informed decision-making (Hosamo et al. 2022). The digital twin concept is also used in the AEC industry for information integration, occupants-centric design, performance monitoring, operation optimization, predictive maintenance, etc.
- **Internet of Things and smart sensors:** The Internet of Things is a vast network of physical objects (called things) embedded with sensors and software for collecting and exchanging data with other devices or systems over the Internet. Sensors in equipment, concrete, or components of infrastructure facilitate their real-time monitoring. Sensors, like smart helmets, watches, vests, etc., are wearable for ensuring workers' safety and tracking their locations. Sensors facilitate real-time monitoring of equipment, workers, and components of infrastructure. Sensors embedded in different components of infrastructure collect real-time data continuously on the infrastructure's health condition, energy use, carbon emission, safety, security, and environmental impacts. These are also used for tracking fluctuations in temperature, humidity, vibrations, energy usage, and structural health. Sensors are used for tracking site dynamics, like performance, quality, logistics, etc.,
- **Artificial intelligence and machine learning:** artificial intelligence simulates human intelligence and machine learning involves developing a program or algorithm that learns from data to make predictions. These enhance image recognition for construction quality checks, safety monitoring by processing imagery obtained from drones, and decision-making through data analytics. Artificial intelligence and big data facilitate the development of optimal architectural plans, cost forecasting, predictive project planning, scheduling, site layouts,

maintenance plans, and risk management. Research in machine learning will improve design, site safety, cost estimation, scheduling, and O&M by analyzing data for predictive insights,

- **Robotics and automation:** is used in autonomous excavators, autonomous bulldozers, automated construction processes, brick-laying robots, 3D printing using robotic arms, etc. Automation is used to perform repetitive, heavy, or tedious tasks, like bricklaying, concrete pouring, welding, lifting of heavy components, etc. Robotics and automation enhance precision, speed up operations on sites, and reduce reliance on manual labor in risky working conditions.
- **Drones and unmanned aerial vehicles:** are used for surveying, site mapping, topographical mapping, and taking imagery used in construction progress monitoring, safety inspection, etc. Drones mainly provide high-resolution imagery that reduces manual surveying time and human risk in inspecting inaccessible areas.
- **Augmented reality, virtual reality, and mixed reality:** augmented reality is the technology that overlays digital information and virtual objects into the real world and in real-time. In the AEC industry, it is used for overlaying digital models onto a physical site. On the other hand, virtual reality replaces the real world completely with a digital one. It is used in design reviews, immersive client presentations, worker training, etc. Mixed reality is used for integration and interactive collaboration between physical and virtual worlds.
- **Cloud computing and collaboration platforms:** facilitate centralized storage of project data accessible in real-time. Cloud-based platforms are used for sharing BIM models, project-specific data, real-time updates, and performing various analyses. This enhances communication across architects, engineers, contractors, and owners, reducing errors caused by fragmented data.
- **Blockchain and smart contracts:** blockchain is a sort of digital record book that everyone in a group shares. Whatever contents are in the record book, no one can change them, and everyone agrees with it. It ensures transparent contract management for secure and tamper-proof contracts. It secures records of the supply chain, materials procurements, payments made, and reduces disputes between contractors and clients. It helps in automating the payment process when project milestones are reached.
- **3D printing and prefabrication:** speed up the construction stage by printing structures or their components and reducing waste through precise materials use. Large-scale 3D printers for the construction of houses, bridges, and components of infrastructure are being matured. Modular and off-site prefabrication facilitate faster, quicker assembly, and cost-effective construction to support sustainable and low-cost housing.
- **Big data and analytics**: collect and process large volumes of project or site-specific data obtained from sensors, Internet of Things, GIS, BIM, and help in decision-making, productivity tracking, productivity enhancement, detecting inefficiencies, and risk analysis.

9.5 Technological Research Priorities

The research directions to extend the use of digital technologies in different lifecycle stages of infrastructure projects are provided in the following sections:

- **Interoperability:** Interoperability remains a major challenge in the AEC industry, as data exchange among BIM, GIS, Internet of Things, and lifecycle management systems is often fragmented. Future research in Construction 4.0 must focus on enhancing interoperability and integration among BIM, GIS, Internet of Things, digital twins, and lifecycle management systems to create seamless data-driven systems across different lifecycle stages of infrastructure. Future research on BIM and digital twins should emphasize seamless integration across platforms and disciplines. Research is needed to establish common standards and protocols that allow continuous data sharing throughout the lifecycle of a project. The standardization of data exchange formats, as well as the creation of cloud-based platforms and middleware solutions that allow smooth data flow across digital environments, is fundamental for bridging gaps between different lifecycle stages of infrastructure projects.
- **Digital twin:** Another promising research direction is the development of a real-time and self-updating digital twin through Internet of Things sensor data and artificial intelligence-enabled analytics that can reflect the as-built and real-time conditions of infrastructure. Artificial intelligence-enhanced BIM models can be investigated to automatically detect design clashes, optimize layouts, and predict costs, thereby reducing errors and improving project efficiency.
- **Internet of Things and smart sensors**: The potential of Internet of Things and smart sensors in the AEC industry lies in their ability to provide real-time insights into infrastructure; however, several challenges remain unresolved. Future research should focus on the design of low-cost, energy-efficient, and durable sensors that can withstand harsh construction, operational, and environmental conditions. Wireless sensor networks that can cover large and complex construction sites also require further study, especially in terms of connectivity, reliability, and durability. Another critical area is data fusion, where information from diverse Internet of Things devices can be combined with BIM and digital twins to support better decision-making, predictive maintenance, and proactive safety management.
- **Artificial intelligence and machine learning**: Artificial intelligence and machine learning offer enormous potential in predictive and generative applications, but more research is required to realize their full benefits in the construction industry. Predictive analytics can be expanded to identify risks of delays, cost overruns, or safety incidents, allowing managers to act before problems escalate, etc. Another research area is generative design, where artificial intelligence systems propose optimized layouts, suitable materials, and methods that can balance sustainability, cost, and structural performance. In addition, computer vision-powered artificial intelligence can be developed further to automate defect detection, quality control, and safety inspections using images and videos captured from drones or

site cameras. From a research perspective, current findings are largely based on literature reviews and descriptive case analyses, which provide valuable insights into the contextual factors, challenges, and opportunities associated with artificial intelligence adoption. However, to strengthen the robustness and generalizability of these insights, future research should incorporate empirical methods, including surveys, controlled experiments, and statistical benchmarking. Such approaches would offer more comprehensive evidence on how generative artificial intelligence, ChatGPT, and large language models can transform the AEC industry. Ultimately, the careful integration of artificial intelligence not only advances education and workforce training but also enhances work efficiency, safety at the site, and sustainability, preparing the AEC industry for competitiveness.

- **Robotics and automation:** robotics and automation research in construction must focus on adaptability and human–robot collaboration. Current robotic systems often face limitations in unstructured construction sites; hence, future work should explore advanced mobility, perception, and decision-making capabilities. Human–robot collaboration is another promising area, where robots can handle repetitive or hazardous tasks while humans manage complex and judgment-driven operations for improving site safety and productivity.
- **3D Printing and prefabrication**: 3D printing should be further developed to enable the construction of large-scale infrastructure projects using sustainable, locally available materials, ultimately reducing construction costs and environmental impacts.
- **Drones and unmanned aerial vehicles**: are already widely used for surveying and inspection, but future research can greatly expand their scope. Autonomous drone fleets that coordinate with each other to map, monitor, and report on construction sites could significantly enhance efficiency. Artificial intelligence-based image processing and computer vision need to be advanced for automated defect detection, structural inspections, and progress validation. Integrating drone data directly into BIM and cloud platforms also presents an important research area. It will enable continuous and real-time site documentation that will help in developing a decision support system that may work without manual intervention.
- **Augmented reality, virtual reality, and mixed reality**: research into augmented reality, virtual reality, and mixed reality should prioritize usability and affordability for large-scale adoption in the AEC industry (OpenAI 2025). Current wearable devices being used in the industry are expensive or cumbersome, making their use on-site a challenging task. Future studies should focus on developing lightweight and cost-effective augmented reality glasses and devices that seamlessly overlay BIM models onto physical environments for real-time guidance. Similarly, augmented reality-guided workflows must be enhanced to reduce errors during assembly and installation processes. Virtual reality has great potential in training simulations, enabling workers to practice safety procedures, equipment handling, and emergency responses in immersive and risky environments. Mixed reality should also be explored further for interactive collaboration between remote teams.

9.6 Conclusions

GIS is the technology that is not generally associated with the AEC industry; therefore, it is a little difficult to implement immediately in the AEC industry. Several GIS-based software tools are suggested in the literature and their applicability on infrastructure projects has been demonstrated through case studies; however, their implementation again on real-life projects is rare. The professionals involved in the lifecycle stages of infrastructure projects do not find it time and cost-effective to invest in the GIS. Hence, a holistic approach is required where GIS is conceived, redesigned, and used in all lifecycle stages of infrastructure projects. Enhancing the implementation of GIS-based software tools in practice in the AEC industry needs further consideration.

The circular economy principles support waste management from the earliest design stages, ensuring that sustainability is not just an afterthought but a core principle throughout the lifecycle. Despite many advantages of the circular economy, its adoption in the AEC industry faces several barriers, including the lack of supportive policies, market resistance to secondary materials, limited awareness among stakeholders, and insufficient infrastructure for large-scale recycling. However, as environmental concerns intensify and raw material scarcity increases, the circular economy is gaining momentum as both a sustainable necessity and an economic opportunity. By embedding the circular economy into design, construction, and demolition stages, the AEC industry may shift toward a future that prioritizes resource efficiency, waste minimization, and environmental responsibility. In this context, the circular economy is not just a sustainability framework but a pathway to redefining how infrastructure is built, maintained, and renewed in a way that aligns with global goals for climate action and sustainable development.

In Construction 4.0, multiple digital technologies, like GIS, BIM, digital twins, Internet of Things, artificial intelligence, augmented reality, virtual reality, mixed reality, 3D printing, robotics, and blockchain, are transforming the way infrastructure projects are designed, planned, constructed, maintained, and managed. However, to achieve the full benefits of recent digital technologies, significant research and development are still required. The research on the widely used digital platform BIM must focus on improving interoperability with other required digital platforms, as discussed earlier. In order to provide a seamless data-driven platform across all lifecycle stages of infrastructure projects, future research must focus on enhancing interoperability and integration among BIM, GIS, Internet of Things, and digital twins. This will also help in the full automation of the updating process in BIM through real-time data received from the Internet of Things. The updated BIM will play a creative role in the planning and management of lifecycle and predictive maintenance.

The research in the field of GIS has been discussed in detail. Furthermore, research in GIS must emphasize integration with BIM and digital twins to support smart geospatial decision-making and sustainability analysis. In the field of digital twins, research must focus on developing accurate and high-fidelity replicas of infrastructure that are synchronized with data obtained from the Internet of Things. The research on producing affordable sensors for real-time monitoring will improve real-time updating. For real-time data transfer between physical infrastructure and digital twins, research is required to design secure, energy-efficient, and economical sensors for their large-scale deployment and to ensure secure data transmission. For monitoring workers on sites, equipment operations, ensuring safety, and ensuring good working conditions, interest is also growing in integrating the Internet of Things with BIM, GIS, and digital twins.

Artificial intelligence can greatly enhance the capabilities of GIS, but its applications in the construction industry are still emerging. Research is required to develop artificial intelligence-enhanced GIS software tools that can automate mapping, terrain analysis, site selection, and risk prediction. Machine learning techniques can also be explored for using massive datasets from drones, satellites, and sensors to provide automated insights into site conditions, construction progress, quality assessment, safety issues, etc. Integrating artificial intelligence with GIS could reduce reliance on manual data interpretation and accelerate decision-making for large and complex projects. Artificial intelligence-enhanced GIS could predict environmental impacts and support smart development.

When integrated with the Internet of Things and digital twin, GIS could enable real-time monitoring of environmental impacts during construction and O&M stages. This would help project teams to adopt green construction practices and ensure compliance with sustainability standards, making GIS an essential tool in developing climate-resilient infrastructure. The fusion of GIS with the Internet of Things and digital Twins is a main research area that has significant potential in the AEC industry. GIS provides the spatial framework, while the Internet of Things devices capture real-time data on equipment usage, site conditions, and environmental factors. Future research should develop GIS-enabled digital twins that integrate geospatial, design, and operational data in real time. Such systems will support dynamic monitoring of construction sites, predictive maintenance of infrastructure, and proactive risk management, enabling smarter decision-making across the lifecycle stages of infrastructure.

Artificial intelligence, the Internet of Things, and GIS can be used to monitor carbon emissions and optimize resource use in real-time on sites. These digital platforms integrated with BIM and blockchain could support material reuse and recycling by maintaining detailed inventories of materials used in different components of infrastructure throughout their lifecycle. Another crucial research direction is embedding life-cycle assessment tools into digital twin models, allowing designers, constructors, and operators to evaluate the long-term environmental impact of their projects and adopt greener solutions.

The research in the area of artificial intelligence and machine learning will emphasize predictive analytics for further improvements in intelligent decision-making through big data. Further research is needed in the areas of developing data management standards, cybersecurity, and predictive modeling. Similarly, augmented reality and virtual reality applications have to be fully explored in design visualization, immersive training, real-time collaboration, and safety training. Future research in the area of robotics and 3D printing needs to improve materials use and the construction of complex structures. 3D printing research must focus on exploring the use of more durable and sustainable materials, ensuring the structural integrity of printed components of infrastructure, and combining automation and robotics for efficient on-site applications. Research in robotics and automation must explore advanced autonomous machinery for the AEC industry. The human–robot interaction on construction sites also remains a major research area.

The blockchain research must explore its role in creating more secure contract management, more transparency in the supply chain, and safeguarding data in collaborative digital systems. Data security and scalability should be the focus of blockchain and smart contract research. Smart contracts should automate the payment process based on predefined milestones, procurements, and dispute resolution. Further research should also investigate blockchain for supply chain transparency, enabling better tracking of materials, reducing fraud, and supporting sustainable construction. The applications of blockchain in construction are still in their beginning; future research must explore their potential for more transparency and trust. Scalable blockchain frameworks must be developed to handle the complexity and size of modern infrastructure projects.

Sustainability always remains the core research area in Construction 4.0; digital technologies provide new opportunities to advance this area. Research on sustainability and circular economy practices is essential to make infrastructure intelligent, sustainable, and green in the future, and help in minimizing carbon emissions, reusing materials, and prolonging lifecycles. Overall, research in Construction 4.0 technologies revolves around integration, automation, sustainability, standardization, and data security. The convergence of these technologies requires interdisciplinary research to develop a fully connected, intelligent, energy-efficient, and sustainable lifecycle.

Despite its potential as discussed in previous chapters, GIS adoption in the AEC industry remains limited. Research should investigate ways to design simplified GIS platforms tailored to architects, engineers, and contractors, who may not have advanced geospatial training. Moreover, developing education and training programs is essential in building GIS literacy in the AEC industry. Studies on organizational and cultural barriers in the adoption of GIS may also help in formulating strategies to encourage its wider industry use. Without improving usability and awareness, the benefits of GIS in industry will remain underutilized.

Age, computer understanding, experience, and education level of professionals in the AEC industry are factors that may influence the implementation of GIS technology in the lifecycle stages of infrastructure projects. Scenario planning shows that widespread adoption of artificial intelligence can unify education, training, and practice, fostering sustained growth, innovation, and competitiveness in the industry. While artificial intelligence tools demonstrate strong potential for large-scale infrastructure projects, adoption in small and medium-sized projects faces challenges with respect to cost, ease of integration, and artificial intelligence literacy of the workforce. Cultural issues may be another major barrier. By understanding how an individual in the AEC industry may resist the implementation of GIS technology, the resistance to the adoption may be better accommodated through research and development. This, in turn, will enhance global competitiveness and enable researchers and practitioners to understand how GIS should be introduced in the AEC industry. To stay competitive in the long run, such evolutionary changes must be introduced in the AEC industry. Although the book covers reported application areas of GIS, BIM, and integration of BIM and GIS in the lifecycle stages of infrastructure projects, their successful implementation in real-life projects will depend upon future practices and research. Future research may also help researchers and practitioners understand market needs and enable them to understand how GIS may be introduced. The research community, software developers, and construction professionals together may bring GIS as a huge opportunity in the different lifecycle stages of infrastructure projects in the AEC industry.

9.7 Exercises

Question 9.1: What are the main challenges AEC industry professionals face due to the use of different software tools and formats for project information?

Question 9.2: Why is interoperability considered a critical issue in the lifecycle stages of infrastructure projects? How does BIM help in overcoming the fragmented nature of the AEC industry?

Question 9.3: Why is research on the integration of BIM and GIS still considered essential for future infrastructure development?

Question 9.4: What limitations exist in the 3D modeling capabilities of GIS compared to BIM?

Question 9.5: What are the challenges of developing 3D topological models in GIS, and why are they important in planning?

Question 9.6: What role does the Open Geospatial Consortium play in addressing interoperability issues between BIM and GIS?

Question 9.7: Why is operational-level visualization in GIS important for construction planning, and what additional capabilities are needed to achieve it?

Question 9.8: Explain how web/server-based GIS improves collaboration and information sharing in infrastructure projects.

Question 9.9: What do the terms Construction 1.0, 2.0, 3.0, and 4.0 mean in the AEC industry?

Question 9.10: How did the use of technology transform the AEC industry from manual work to digital approaches?

Question 9.11: What role does sustainability play in Construction 4.0?

Question 9.12: What are the various technologies that drive Construction 4.0?

References

Akhimien, N. G., Latif, E., and Hou, S. S. (2021). "Application of circular economy principles in buildings: A systematic review." *Journal of Building Engineering*, 38, 102041.

Akınci, B., Karımi, H. A., Pradhan, A., Wu, C. C., and Fichtl, G. (2010). "CAD and GIS interoperability through semantic web services." In: H. A. Karimi and B. Akinci (eds.), *CAD and GIS Integration*. Boca Raton: Auerbach Publications, Taylor and Francis Group.

Bansal, V. K. (2011). "Use of GIS and topology in the identification and resolution of space conflicts." *Journal of Computing in Civil. Engineering*, 25(2), 159–171.

Bansal, V. K. (2012). "Application areas of GIS in construction projects and future research directions." *International Journal of Construction Management*, 12(4), 17–36.

Bansal, V. K., and Pal, M. (2008). "Generating, evaluating, and visualizing construction schedule with geographic information systems." *Journal of Computing in Civil Engineering*, 22(4), 233–242.

Borrmann, A., and Rank, E. (2009). "Topological analysis of 3D building models using a spatial query language." *Advanced Engineering Informatics*, 23, 370–385.

Hosamo, H. H., Imran, A., Cardenas-Cartagena, J., Svennevig, P. R., Svidt, K., and Nielsen, H. K. (2022). "A review of the digital twin technology in the aec-fm industry." *Advances in Civil Engineering*, 2022(1), 2185170.

Jelodar, M. B. (2025). "Generative AI, large language models, and ChatGPT in construction education, training, and practice." *Buildings*, 15(6), 933.

OpenAI (2025). ChatGPT (Version 5.1) [Large language model]. https://chat.openai.com/

Zhu, J., Wang, X., Chen, M., Wu, P., and Kim, M. J. (2019). "Integration of BIM and GIS: Geometry from IFC to shapefile using open-source technology." *Automation in Construction*, 102, 105–119.

Index

For Product Safety Concerns and Information please contact our
EU representative GPSR@taylorandfrancis.com Taylor & Francis
Verlag GmbH, Kaufingerstraße 24, 80331 München, Germany